SHARK DOWN UNDER

Shark Attacks in Australian Waters

SHARK DOWN UNDER
Shark Attacks in Australian Waters

Alan Sharpe

Other books by Alan Sharpe

Years of Change
Historic Australian Crimes
Colonial New South Wales
Bushranger Country
50 Crimes That Shocked Australia
Manly to Palm Beach
Streets of Old Sydney

Kingsclear Books
(Imprint of Atrand Pty Ltd
ACN 001 904 034)
Suite 2, 77 Willoughby Road
Crows Nest, 2065
Phone (02) 439-5093, (02) 436-2576
Facsimile (02) 906-2035
ISBN 908272-33-2

Designed by Michelle Havenstein
Printed by Griffin Paperbacks

Photo credits:
Front Cover
Top: Barry Smith
Bottom: Sydney Aquarium
Back Cover: Paul Nevin

Foreword

"The only safe precaution against shark attack is to bathe in the protected areas. Attacks cannot be anticipated, and their likelihood does not vary with the time of the year or locality."

Gilbert P. *Whitley*

This book is written for all Australians who love to get down to the beaches, rivers and harbours at weekends or holidays, to cool off and forget the cares and pressures of living.

There is no story like a shark story and inside our heads is a jumble of facts and fallacies, half-truths and exaggerations. Somewhere in our minds too, are the newspaper reports of shark attacks. The more horrific ones stand out most.

The shark fatality which made the deepest impression on me was the tragic death of the young Sydney actress, Marcia Hathaway in Middle Harbour in January 1963. Until then, I fully believed, as many Australians still do, that a shark would never attack unless you were at least chest-deep. Miss Hathaway was mauled in a mere 75 centimetres of water.

In researching among the many books and feature articles written on the subject I was struck by the differing opinions held among the experts. This is quite understandable. After all, coordinated study of sharks is little more than a generation old and, although our knowledge increases every day, some 250 species of shark inhabit this globe and their behaviour is as complex as that of human beings. I have tried to reach a balanced view of all opinions.

The shark can be likened to that mild-mannered neighbour across the street who has been going quietly about his business for years until one day, suddenly, unaccountably he runs amok with a rifle, killing and wounding. You just never know with a shark.

Human folly persists, especially in the young. A warning automatically becomes a challenge. Many of the reports in this book read like gothic horror stories. The unhappiness and often the self-recrimination of loved ones left behind, can never be chronicled.

Sharks have existed for a period of time almost incomprehensible to the human brain. Their place in natural evolution is established and essential. Sharks are not cruel or vindictive. They are motivated by their ceaseless quest for food. In no way do they regard humans as a natural enemy, although, sadly, many humans see all sharks as such.

The purpose of this book is to help to get the facts right; to untangle some of the lines on this fascinating subject; to warn, to advise and, at the same time, hopefully to interest you.

Contents

CHAPTER ONE

▲▲▲▲▲▲▲▲▲▲▲▲▲

One in a million chance

The early 1950s saw the start of an unprecedented invasion of the marine environment by a creature who had hitherto kept largely to the land. The invention of the self-contained underwater breathing apparatus (scuba) allowed human beings to dive deeper and remain underwater for long periods. In entering the terrain of the shark in increasing numbers we exposed ourselves more and more to the risk of attack.

Of all the predators of the sea the most fearsome is probably the white pointer, or great white. Our coastal waters are patrolled by one of the world's largest populations of great white sharks. It is quite surprising to note, then, that before June 1993 there had been only three attacks by these awesome creatures on divers in Australian waters. The chances of two such attacks in any one week were phenomenally small. In June 1993 the incredible odds came up in Tasmania and New South Wales within four days.

TASMANIA

Therese Cartwright was an experienced scuba diver. In 12 years she had logged over 200 dives. However, on Saturday 5 June 1993, as she jumped from a boat into the clear cold water near Tenth Island, about 6.5 kilometres off the north-eastern coast of Tasmania, she had not been diving for well over a year. Terry, 34, had been busy with her master's degree in nursing at the University of Tasmania and with her children, 16-month-old Paul and six-year-old quadruplets, Sarah, Thomas, James and Luke.

Therese and her husband Ian had hit the headlines in Australia with the birth of the quads, the result of fertility drugs. That day Therese was once again to make front page news, this time on a far from joyous occasion.

The small rocky Tenth Island is home to the Barrenjoey fur seal colony. It is one of Australia's finest locations for observing seals in the water. On the way out there was some discussion amongst the dive party about the possibility of encountering sharks, attracted by the seals at play. The risk was dismissed. As Terry's brother, Philip Edwards, was later to say, 'Tasmania is not renowned for shark attacks. It's one in a million.'

The dive boat anchored about 100 metres from the island. It was a glorious sunny winter's day and diving conditions were perfect, the water calm and clear. The Cartwrights and the four other members of the party agreed to dive in two groups of three. One group would take care of the children on the boat while the others dived.

First into the water was Steve Eayrs, who had organised the outing. Terry jumped in next. Despite her 7 millimetre dark blue wetsuit she was heard to exclaim at the cold as she hit the water. Jo Osborne followed and the three began to kick along the surface towards the island. Terry, adjusting her gear, fell a little behind and Ian Cartwright called to her from the boat to keep up with the others.

Terry reached the others and the three signalled to each other that they were ready to go down. It was 10.55 am. Letting the air out of their buoyancy vests they began the descent, Jo and Steve quickly reaching the ocean floor, about 8 metres down. At about 4 metres Terry stopped, apparently having difficulty in clearing her ears. Steve looked up at her. A big white pointer, about 4 to 5 metres long, had hold of her and was shaking and bumping her violently. Jo looked up to see what Steve was looking at. She could see Terry's silhouette against the surface of the water as the shark moved back and then went in again for the kill. As it turned and swam swiftly away the pair saw Terry's yellow fin sticking out of its mouth.

Terrified, Steve and Jo swam to a hollow in the rocks about 14 metres deep. Here they hid for a while before making their way to the island where they surfaced and waved for help. Ian Cartwright and another member of the party took the boat's dinghy over to the island. Cartwright was clearly worried.

'Where is she?' he wanted to know. 'Look mate, I think a shark's got her,' Steve told him. 'There was a very bloody big shark there. I think she's gone.'

Cartwright began to shout and scream in shock and grief. Once back on the boat he wanted to be alone. He took charge, radioing the police and then taking control of the boat. As he skippered the boat back to port a grisly report came over the radio. A wetsuited human leg, cleanly severed at the top of the thigh and still shod with a yellow diving fin, had been picked up by the Launceston Port Authority at 3.30 pm. Nothing more was found of Therese Cartwright.

Therese and Ian Cartwright when the quads were first born.

Andrew Fox, an Adelaide marine biologist and son of Rodney Fox (who survived a white pointer attack in 1963) theorised that the shark may have come from below, seeing Therese Cartwright as a 'silhouette target'. It is possible that she may have been mistaken for a seal in her dark wetsuit.

According to Fox a white shark normally cruises at 3 kilometres per hour but accelerates to 16 kilometres per hour when it is about to make an attack. It then sinks its bottom jaw into its victim, using it as a holding brace while its wider top teeth act like a saw, and will then swim and shake its prey violently from side to side. Although white sharks are not fussy about what they eat, including other sharks, they do not often eat humans and many people consider that attacks on people are cases of mistaken identity.

BYRON BAY, NSW

In 1982 surfer Marty Ford was killed by a shark at Byron Bay on the far northern coast of NSW. The most easterly point of Australia, Byron is a haven for surfers and other tourists. Julian Rocks, a marine reserve 3 kilometres off Cape Byron, attracts around 50,000 divers each year. In the local pubs patrons occasionally sing a ditty based on a popular advertising campaign, 'Have you eaten a Ford lately?' This gruesome piece of bad taste is based on an incident that took place 11 years after Marty Ford's death, when the next shark attack took place in the waters of Byron Bay.

Early on the morning of Wednesday 9 June 1993 honeymooning couple John and Debbie Ford from Sydney, both experienced divers, boarded a dive boat from the Sundive shop with three other divers and were taken to a popular site in open waters known as Spot X.

Four days earlier newspapers had carried the story of a fatal attack on a diver in Tasmania by a white pointer. Therese Cartwright had been diving in a seal colony. Although sharks were regularly seen by fishermen in the area, the Fords were not particularly worried. White pointers are extremely rare as far north as Byron Bay, preferring the cooler southern waters. In fact one of the objectives of their dive was to observe grey nurse sharks which frequent the waters at this location.

Shortly before the party ran out of air a large white pointer was sighted. At 9.30 am the five divers ascended, holding on to the boat's anchor line. Three metres from the surface they made a safety stop. As they waited a 5.5 metre white pointer appeared beneath them and moved towards Debbie. John, pushing Debbie out of the way, placed his own body between the shark and his bride. The next moment John had disappeared into the jaws of the shark.

THE DAILY

Telegraph Mirror

SYDNEY, Friday, June 11, 1993 | WEATHER: Rain developing, 15 degrees | Phone: 288 3000 | 60 cents*

I LOVED HIM

'He took his own life for me'

Jan Holm hugs her weeping daughter Debbie Ford yesterday as Debbie spoke of her 'hero and good husband'. Picture: PAUL RILEY

John Ford . . . tribute

By MARK JONES in Byron Bay

NEWLYWED Debbie Ford yesterday held a memorial service at the spot where her husband sacrificed his life to save her from the jaws of a white pointer shark.

Shattered at losing him just 15 days after their wedding, Mrs Ford returned by boat to the Byron Bay diving site where her husband, John, was killed by the 5m shark on Wednesday.

The grieving wife said goodbye to the man she loved. Her hero.

"He took his own life for me," Mrs Ford sobbed.

"He was a hero and a good husband. I loved him dearly."

Her father, Kevin Holm, said: "It was a very private service. A chance for my daughter to say goodbye to her husband."

As the hunt for the killer continued off the NSW far-north coastal resort, Mrs Ford, 29, prayed and placed a wreath on the ocean.

She wept as she put the floral tribute and her husband's boating cap on the calm blue water where her 31-year-old husband died after pushing her from the path of the shark.

She was comforted by a Catholic priest who conducted the private service aboard a rescue boat.

Earlier, Mrs Ford broke down as

Continued Page 4
Editorial — Page 10

TV: P103; Business: P30; Comics: P88; Crosswords: P87; Lottery (5017): P90

TeleClassifieds P43 (288 2000)

'I loved him,' the Telegraph Mirror front page the day after John Ford's memorial service, three weeks after he was married.

Debbie was taken from the water in hysterics. Local dive boats, sea-rescue crews and fishermen immediately set out in pursuit of the shark. Two Byron Bay fishermen who rushed straight to the scene found the shark circling a dive boat. It was still hungry and snapped up their bait as soon as they dropped it in the water. For the next 90 minutes the two men underwent a terrifying ordeal as the huge beast, almost as long as their 6 metre metal boat, dragged them around while they tried to get it close enough to the boat to shoot it with a .303 rifle. During this time the shark regurgitated the mutilated torso of John Ford, still wearing a shredded portion of his wetsuit. Eventually it bit through the line and got away.

The hunt for the shark continued over the following days, amid much controversy. The curator of Sea World on Queensland's Gold Coast, Mark Smith, said it was unusual for a white pointer to attack divers but that it was an alarmist and unjustified action to hunt it down. His opinion was supported by a number of marine scientists. However, a local surfer was quoted in the press as saying that 'catching it will be like laying a ghost to rest' and Ron Boggis, one of the two fishermen who hooked the shark, said, 'We want to get it bloody bad for the sake of that poor woman'. The search continued for two days during which time some of John Ford's wetsuit and his fins, tank and weight belt were recovered. The shark was not found. Over the next weeks various parts of Ford's body, including his head, were found on the sea bottom.

The theory that sharks will not attack humans when they are well fed does not hold up in this case. The area was teeming with mullet and 18 tonnes had been caught that week.

CHAPTER TWO

Dangerous shallows

There is an unreality about the month of January in Australia. The race to Christmas, those frantic last hours, have slowed to a shuffle. Half the country is on holiday. The other half find themselves in a kind of vacuum, subdued by the glaring blue sky. The issues of the year ended have lost all significance. The world outside is blurred in the shimmering heat.

Attuned to this universal apathy, newspapers cast about for themes which will catch their readers' flagging attention. Sharks are always a good bet: sightings, warnings, near misses. Everyone seems to be at the beach. It's shark time again. The front page headline has sometimes carried the rare story of a particularly gruesome shark attack; even more rarely has the front page headline been the subject of a warning. This happened on page one of the Sydney Morning Herald, Australia Day, Monday 28 January 1963:

AUSTRALIA DAY SURFERS WARNED OF SHARKS

The beach patrol aircraft had counted 66 sharks on the previous day's flight along Sydney's northern beaches. From Collaroy in the north to Wanda in the south, 12 beaches were closed on and off during that weekend. A school of nine hammerhead were sighted at Palm Beach the previous afternoon. A shark was spotted less than 46 metres from swimmers at Maroubra.

Australia Day Monday was the kind of day Sydneysiders call 'good shark weather.' It was a warm 26°C with a high humidity. Threatening clouds hung motionless in the sky. The seas were dark and calm. A motor cruiser moved slowly through the waters of Sydney's broad Middle Harbour. On deck, picnic party of six sprawled on deck in their swimsuits, grateful for the breath of air.

The boat turned into Sugarloaf Bay, a ragged fringe of sand

just north of Balmoral Beach. Two of the party went off to forage for oysters off the rocks. Another couple, an attractive young woman and her fiancé, happily cooled off together in the shallows. In a mere 75 centimetres they dunked themselves and swam briefly and laughed to each other. Perhaps it was a sixth sense, or even the *Herald* headline that morning, that prompted journalist Frederick Knight to warn his fiancé, Marcia Hathaway, a Sydney

Bystanders strain every ounce of muscle to push the stalled ambulance carrying Marcia Hathaway up the steep grade from the boatshed.

actress, not to go out too far. The water was lapping at her hips when she suddenly screamed out, 'I've been bitten by an octopus!'

Knight was a couple of arm-lengths from her when the shark seized her right leg, below the calf. In the terrible minutes that followed a grim struggle took place between man and beast for the possession of 32-year-old Marcia Hathaway. 'At one stage I had my foot in its mouth,' Knight recalled later. 'It felt soft and

spongy.' In its second lunge, the shark embedded its teeth into her upper thigh.

J. Delmage, a member of the party, rushed to assist Knight. Together they succeeded in releasing the girl's torn leg from the shark's jaws. Sheets from the cabin bunks were ripped to make a tourniquet and the cruiser, with its stunned passengers, moved across to the boatshed at the foot of Castlecrag. Without hesitating Fred Knight dived overboard and swam 20 metres to a house at the water's edge, telling the occupants to ring for the ambulance. He swam back and the boat made its way to Mowbray Point.

The white-faced girl swathed in make-shift bandages, opened her eyes briefly and reached out her hand to her anguished fiancé. 'Don't worry dear,' she whispered, 'God will look after me.'

The Central District Ambulance was already waiting. Marcia Hathaway was barely conscious. The ambulance officers tried oxygen to revive her. Time was running short. But fate was not on the side of Marcia Hathaway on that day. The steep grade leading from the water was slippery. The wheels of the ambulance skidded as the vehicle strained forward a few feet. The driver pumped the accelerator. The tyres screeched. Then the clutch went; burnt out. An incredible scene followed. Thirty able-bodied men closed in around the vehicle and with every ounce of muscle, heaved and strained to push it up the hill. The heavy ambulance refused to move.

Now the distant howl of a siren could be heard. It grew louder and a second ambulance appeared at the crest. The unconscious girl was hastily transferred and moments later the ambulance was shrieking through several kilometres of streets to the Mater Misericordiae hospital. The ambulance men tried everything to revive her. It was already too late. Marcia Hathaway was dead on arrival, just 20 minutes after the attack.

Following the tragic death of Marcia Hathaway on 28 January 1963 the hunt for a killer shark was again on in Middle Harbour. Boats were joined by aircraft but the murky water concealed any lurking predator from the air. TV entertainer and big game fisherman, Bob Dyer, refused to take part in the hunt. 'The more people who go up there putting out horse meat and legs of lamb, the more sharks it is going to attract.'

From fragments of teeth in Miss Hathaway's wounds it was believed a bronze whaler was responsible. The fact that a group of lads inspecting the scene of the attack were startled when a bronze whaler surfaced close by them, seemed to confirm this. Then Tom McCulla, the fisherman who caught the shark which took 13-year-old Ken Murray three years earlier, announced he had caught a bronze whaler in company with Mike Campbell, at 10 pm the next day, at a spot 90 metres from where Marcia Hathaway was attacked.

'We skull-dragged the shark for about an hour until we beached it, then gave it a couple of dongs on the head with an oar. It weighed 600 pounds (272 kilograms). We loaded it in Mike's station wagon and took it home.'

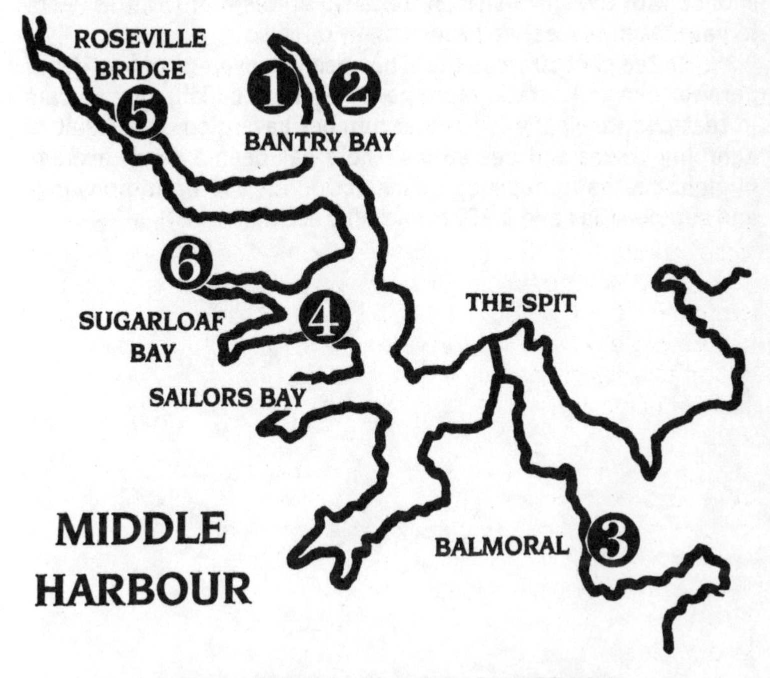

FATAL ATTACKS IN MIDDLE HARBOUR

(1) *Zita Steadman*, 1942; (2) *Denise Burch*, 1942; (3) *John Willis*, 1955; (4) *Bruno Rautenberg*, 1955; (5) *Ken Murray*, 1960; (6) *Marcia Hathaway*, 1963.

While neighbours came to stare and shudder at the dead bronze whaler, a service, attended by many stage personalities, was held at St Stephens, Macquarie Street, on Thursday 31

January 1963. It was said by those who knew Marcia Hathaway that she was a most gentle and kind person, a true Christian lady who devoted much of her free time to charity work.

Meantime blood samples were taken from the bronze whaler. They were found to contain no human blood. The killer had got away.

The Australian Shark Attack File, held at Sydney's Taronga Park Zoo Aquarium, lists 491 recorded shark attacks in Australia over 200 years, of which 182 were fatal.

Report of a shark attack has always made good newspaper copy especially in Australia where the so-called silly season corresponds with the shark season. Because of this more attacks seem to have taken place than have actually done so.

In the past 20 years there has been an average of one death per year by shark attack. More people have been killed by spiders in that period. Nearly twice that number have died as a result of lightning strikes and bee stings. There has been a yearly average of eight deaths from scuba diving accidents, 306 from drownings and submersions and 2,979 from traffic accidents.

CHAPTER THREE

Out of the past

Few falsehoods and fallacies surround any living creature as they do the shark. These are born of fear, fascination and sheer ignorance. An air of mystery surrounds this primeval creature. Because of the shark's strange and unpredictable behaviour, races in various parts of the world have evolved their own sharklore dating back to the early Greeks.

Some 63 million years ago dinosaurs disappeared. Ichthyosaurs sank out of sight forever in the ocean depths and the last pterodactyl fell out of the sky. Vast changes took place on land and sea. Thousands of creatures came and went. The shark remained. Due perhaps to its aggressive nature and active make-up it not only survived, it scarcely changed. The shark is prehistoric. It is a living fossil.

'He seems to come from another planet and in fact he does come from another time,' says Jacques Cousteau.

Shark remains dating back to the Devonian Age, 50 to 300 million years ago, were found on the banks of a river near Cleveland, Ohio, itself located on the south shore of Lake Erie, in the middle of last century. Today's sharks have the same skeletons of cartilage, pointed snouts and mouths on the underside of their heads. Cartilage is the substance which in humans forms the framework of the ears and nose. There are no bones in the skeleton of a shark. Only its teeth lend themselves to fossilisation.

In 1853 a railway surveyor discovered fossilised shark teeth on a hillside in California - 160 kilometres from the sea! The hill, located 11 kilometres north-east of the town of Bakersfield became known as Sharktooth Hill.

The earliest pictorial record of sharks appear on a Greek vase dating back to 725 BC The scene depicts shipwrecked sailors, one of whom is being devoured by a large fish.

In Australia, Aboriginal artists also chose the shark as a subject. Theo Brown tells of finding one such representation as a boy in the bush above Middle Harbour, close to the present day Wakehurst Parkway. On a large expanse of flat rock he saw a carving which illustrated a huge open-mouthed shark about to seize a female figure whose body was already marked with jagged gashes. The earliest written record of shark attacks came from the pen of the Greek historian, Herodotus, who records that in 492 BC sharks devoured shipwrecked sailors of the Persian fleet. It also turns out that the biblical whale in whose belly Jonah was supposed to have spent three days before popping out alive and in one piece, was, in all probability, a lamie: a great white shark. Several writers over the centuries have subscribed to this idea. In the 1611 translation of the bible, the word whale is substituted by a 'great fish.' A mid-18th century writer, Linnaeus, added further proof by giving the description of its teeth, the one positive characteristic in identifying a shark's species.

In the 16th century there was no English word for shark. They were known by the Spanish name of 'tiburon.' One writer described the tiburon as 'a very great fysshe and very quicke and swifte in the water and cruel devourer.' The 'cruelty' label was to stick for centuries. In AD 206 the Greek, Oppianus, wrote a poem about fishing which still exists. It describes a sea creature which is obviously a shark:

And they roved for food with increasing frenzy,
Being always anhungered and never abating
The gluttony of their terrible maw:
For what food should be sufficient to fill
The void of their belly or never to satisfy
And give respite to their insatiable jaw.

The belief that the shark was a symbol of evil prevailed among sailors in the days of sailing ships. They were convinced the shark would follow in the wake of a ship in which people have died or would soon die. In A Sailor's Life Jan de Hartog wrote:

'The focs'le fantasises about whales, however extravagant they may be but always friendly ... the shark on the contrary embodies all that is evil. When a shark is hoisted on board ship as occasionally happens on long voyages, the afterdeck turns into a slaughter house. The men go berserk in a prehistoric orgy of fury and blood against the creature. At this moment the jolly old tars are not loveable. When the orgy is over there is a bewildered sense of shame and they will pretend that the slaughter had a purpose.'

Shark mysticism is strongest in the regions where sharks are most abundant. Pacific Island races, Central American Indians and Malay fishermen all have their shark gods. In Fiji, sharks are still worshipped. Maoris have made gods of sharks with particular emphasis on the bizarre hammerhead. On the island of Mer (Murray Island) in the Torres Strait the Beizam le, or shark god sect, acted as the law enforcement arm of the priesthood, tracking down and punishing transgressors. In Tonga on the night before going out, fishermen refrain from having sexual intercourse with their wives. They must also confess their sins to the chief. Their 'purity' will thus protect them from the shark.

China-based Flying Tiger Squadron manned by American mercenaries used the tiger shark symbol as an ill-omen for Japanese enemy.

There are even shark legends in remote coastal areas of Ireland. In the County of Wexford it was believed that the soul of a local resident had entered the body of shark that haunted the coast and devoured any unfortunate enough to fall into the sea. A strong belief still persists that the sighting of a shark in waters off Ireland is the heralding of a terrible storm.

One ancient tale is still told on the Hawaiian island of Kapaahoo. Whenever the place was visited by a mysterious and handsome stranger, young village girls began to disappear while bathing in the cove. The fishermen held council and decided to keep watch when the girls were swimming. Suddenly a shark appeared and swept in to attack. A score of spears were thrust into its body. The shark dived out of sight leaving behind trails of blood. Lying under a palm tree near the beach, the fishermen came upon the mysterious stranger. He was dying of spear wounds. Slowly his body turned to stone and when he finally expired it had assumed the shape of a shark. That same stone is on display in a museum in Honolulu. Hawaiians still believe the spear is the most effective weapon against the shark. They claim the impact of a bullet is broken by the water. The spear will sink deeper and make a larger wound.

Prewar B-grade jungle movies abound with shark sacrifices with the hero or heroine tied to the stake at low tide. There was an element of truth in this. Under the name of joujou, West African tribes worshipped the shark and were known to present it with a human sacrifice.

In World War II the glamorous Flying Tiger squadron, highly-paid American aviators who flew for the Chinese even before the US entered the war, decorated the nose of their aircraft with shark teeth. The word 'Tiger' referred to the tiger shark and was believed to have a psychological effect on the Japanese who traditionally detest sharks.

CHAPTER FOUR

▲▲▲▲▲▲▲▲▲▲▲▲▲▲

What is a shark?

In the dawn of time nature created a creature perfectly suited to its role in the world: a magnificent marine killing machine.

Even a newly-born shark might attack within seconds of birth, for the watery jungle beneath the ocean is even more savage than the green jungle in the tropic regions on earth. Out of the water, in man's environment, this lethal, graceful, lightning-swift terror of the seas becomes a ridiculous writhing mass of stinking blubber, the only sign of its former power, its awesome rows of teeth. Some 250 species of shark glide gracefully through the silent oceans and rivers which make up 71% of the earth's surface. Only 10% are dangerous to humans.

The shark is a fascinating subject full of unanswered riddles. The stories are endless, the legends unlimited. Slowly and with infinite patience the scientists and the skindivers are narrowing some of the gaps. At times it seems the more one learns about the shark, the less one understands it. Sometimes, with a flick of its tail, it will scoot away from a completely defenceless diver. At other times it will charge with primeval fury at the bars of an underwater cage and bite viciously at the bars. No one has yet been able to establish a behaviour pattern by which its actions and reactions can be predicted.

Shark knowledge is still very new. It wasn't until Christmas Day 1930 that it was realised sharks could inflict injuries on humans without actually using their teeth. James Knight, 49, was swimming in Homebush Bay on the Parramatta River, when he felt a sharp bump on his leg. He thought he had hit a log. It happened a second time. More heavily. To his horror he saw it was shark. He churned the water frantically to scare it off. It made no difference to the shark who closed in and bumped him again, this time on the right leg. Knight started to swim madly but the shark got in one final bump to his hip before he scrambled, bruised and badly shaken, to the bank. In the years ahead as many people would be injured by a shark's bump as by a shark's bite.

In 1926 William Beebe, the American naturalist, wrote in the account of his Zara expedition, 'My book and legend-induced fear of sharks dominated. I saw them as sinuous, crafty, sinister, cruel-mouthed, sneering. When I came at last to know them for harmless scavengers, all these characteristics slipped away, and I saw them as they really are — indolent, awkward, chinless cowards. A ladyfish has a thousand times less weight and double their courage.' In the light of the knowledge of sharks gained in the past 70 years, Beebe's remarks seem oddly naive.

An entirely different view of the shark is expressed by Jacques-Yves Cousteau, the French naval officer and celebrated undersea explorer who invented the Aqualung diving apparatus in 1943. He describes an encounter with the fearsome white pointer. 'His entire form is fluid, weaving from side to side. His head moves slowly from left to right timed to the rhythm of his speed through the water. Only the eye is fixed. Focused on me, circling with the orbit of the head in order not to lose sight of me for a fraction of a second. His silent circling is a ballet of terrifying yet beautiful force.'

The shark is an opportunist. It hunts the old, the weak and the crippled, first. Hereby lies its ecological significance. For in eliminating the misfits from marine society this 'oceanic vacuum cleaner' maintains the quality of the breeding stock.

SIGHT

The shark's senses are highly developed. What it lacks in sight — and even this could well be better than we suspect — it more than makes up for in its ability to pick up vibrations at great distances.

It is still not clear how formidable a shark's sight is. Certainly its sight is superior to that of humans underwater. Sharks spend their lives peering through the gloom. As it gets darker in the depths, nature has provided them with an illuminating mechanism similar to that of cats. Theo Brown, in his book Sharks, *The Silent Savages*, estimates that a shark can see half the distance again as a diver wearing a face mask under the water. Sharks can also see a fair distance out of the water. In Darwin, in the Northern Territory, they have often been seen leaping high in the air to catch flying foxes from branches overhanging the water.

It is generally believed sharks are colour blind. Research into this as well as shark behaviour generally, began in earnest in World War II. In the production of life craft, it was decided that yellow was not the best colour to use. While it may be readily spotted by rescue aircraft, yellow seemed to be more easily espied by sharks. An American, Dr Bombard, who sailed the Atlantic as an endurance test to see whether man can live off the sea, reported sharks grabbing at yellow rubberised cloth attached to the craft while ignoring other colours.

Divers are convinced bright flashing objects, often part of their equipment, attracts sharks. It is also believed that sharks sometimes mistake divers in dark coloured suits for seals.

During World War II the crew of a military aircraft which was ditched in a tropical region of the Pacific took to the water with jackets wearing their flying suits, some of which were green and others orange. When help arrived all those in the orange suits had fallen victim to sharks.

There is also the story of a shot-down US airman in World War II who succeeded in distracting a particularly inquisitive shark, by tearing the white pages from his shark defence manual and sprinkling them over the surface of the sea. The myth that sharks are less likely to attack a coloured swimmer than a white has been proved untrue, as we shall see later.

A number of shark families have a third eyelid, known as a nictitating membrane, which may be pulled up over the eye to protect it as the shark bites. White pointers do not have this membrane but their eyes may roll back in their head as they attack, thus giving a similar appearance of a white eyeball.

SMELL

Sharks have an acute sense of smell. It will attract them from great distances. Not only can they smell fish in a boat, divers are advised to wash their hands after handling fish, before going over the side.

Thor Heyerdahl believes a shark's sense of smell is of prime importance. The crew of the Kon Tiki found dangling their legs over the side a risky business at the best of times. For the most part the string of sharks who constantly followed the raft took little notice. Once blood or offal had been tossed into the sea, the sharks were aroused. If a foot or limb was dangled over the side then, the sharks not only leapt after it, they even sunk their teeth into the logs where the foot had been.

The shark's response to blood has also proved unpredictable. Gallons of animal blood flooded into the sea to attract a killer after an attack, has failed to bring him within coo-ee. Yet a swimmer's cut hand can bring them in like bees around a honeypot. It appears that blood in company with movement, like the wounded fish, is irresistible to the shark.

An incident which illustrates the shark's reaction to human blood occurred off the coast of Cuba in December 1948. Tony Latona, 19, and Bent Jeppsen, 14, were mess boys on the Danish ship *Grete Maersk*. Strolling on the afterdeck after dinner, Tony slipped on the wet deck and slithered overboard. The elder boy screamed 'Man overboard!' but no one appeared to hear. There wasn't a moment to lose. Courageously, Jeppsen kicked off his shoes, grabbed two lifebelts and plunged into what would become a living nightmare.

Miraculously the boys found each other in the darkness. They yelled and screamed only to see the lights of their ship edge further away to be swallowed up by the night. They were alone in the vast, liquid darkness some 15 kilometres from the Cuban coast. A few hours passed. Then the water began to ripple around them. They peered into the blackness and to their terror could make out triangular fins cleaving the water.

Suddenly one of the sharks moved in and either bit or grazed Bent Jeppsen on his left foot, opening two bleeding gashes. The two terrified boys flailed the water with their arms and feet and the sharks momentarily held off. It gave Jeppsen enough time to wriggle out of his pants and bind them around his foot to stop the flow of blood.

All was silent for about an hour. Then the improvised bandage slipped off and the sharks were there again, only this time they did not hesitate. They struck at Jeppsen. He went under still trying to fight them off. Amazingly they entirely ignored the other boy, Tony Latona. No shark came near him that night. At daybreak he could see the coast of Cuba on the horizon. He kept swimming for it but the ebbing tide tugged him back out to sea.

All through the day he tried. Then it was night again. By this time the boy was so exhausted he was indifferent to anything. He remembered only being bumped by a heavy object and the skin of his buttocks being grazed as the seat of his pants were ripped away. A centimetre more would have drawn blood. He would not have lived to tell the tale had he not been picked up by Cuban fishermen next morning.

One intriguing aspect of shark behaviour, difficult to pinpoint, is their response to 'fear odour.' It is believed humans give off a particular odour when they are afraid. It's the theory that this is why some animals, wild and domestic, act friendly towards some people, antagonistic towards others.

'Don't let him see you're afraid,' is the advice often given to those who hesitate before a growling dog. The dog senses he has the upper hand. It might well be he can also smell one's fear. This mental attitude might explain why experienced divers are seldom attacked. The diver may be cautious and watchful. If he were really afraid he would not be entering shark waters in the first place.

TOUCH

A shark will butt or nudge an object that has it puzzled. There are any number of reports of swimmers getting butted, sometimes without realising it was a shark. The shark relies on the tip of its snout or the edge of its fins for touching. Its snout appears to be particularly sensitive. Many times a skindiver has warded off a shark with a smart smack on the snout.

In the early days of experiments with sharks it was found that even though deprived of sight and hearing, the shark continued to respond to disturbances in the water. From head to tail and right around its body the shark has a chord of nerve tissues. This is the 'radar system' that picks up vibrations from great distances. Low-frequency vibrations trigger sensitive hair cells in these fluid-filled canals beneath the skin. If the impulse is regular and rhythmic, the shark may continue on its way. If the movement is erratic, jerky, like the thrashing of a wounded fish or a struggling swimmer, the shark's appetite is whetted and it will come in to investigate.

In an experiment conducted in the waters off Florida, taped sounds of wounded fish and floundering swimmers were played beneath the surface. From a helicopter scientists saw an immediate reaction from the sharks in the vicinity. Some as far as 274 metres swirled about and homed in on the sound. There appears to be little evidence that sharks themselves produce any sound.

HEARING

To the human ear, the undersea world is silent. To the creatures who inhabit the deep, the ocean gives up a whole chorus of sounds. Underwater sound waves travel at about 1.6 kilometres per second, four times faster than air waves. These sounds are transmitted to the shark's brain through its 'ears', two tiny channels in the head, lined with sensitive hairs. Working in conjunction with its information system, the sensitive lateral lines, the shark's ability to locate the source of the sound is highly developed. Such assets are vital to the survival of the ocean's most formidable predator.

In experiments conducted at the Cape Haze Marine Laboratory two sharks were conditioned to push a target which rang a submerged bell. They would then receive food at the location of the bell. Having grasped this association, they were trained to ring the bell in order to get food dropped in another location. During the height of winter the sharks refused food for ten weeks. Yet, in spite of their so-called 'sardine-tin sized' brain, the sharks retained their knowledge over the period and when they were ready to resume eating, they rang the bell again. The fact that they associated the ringing bell rather than the pushed target with food was evidenced by the fact that when a thump at the target failed to produce a ring, they circled and had another go until it did.

Among the various forms of repellents described in a later chapter, those associated with the shark's hearing might one day be found to be the sought-after 100% positive method.

One sense conspicuously lacking in a shark is its ability to feel pain. Cold-blooded animals cannot feel pain. A shark will take an incredible amount of punishment. All kinds of bizarre tales are told of sharks racing off with a harpoon in the back, a spear in the gut or a bullet in the head. The shark's indestructibility is legendary. Hours after it has been landed on the deck it has been known to snap impulsively.

One story is told of a white pointer caught after a one hour tussle. Eventually it had to be shot through the head. A shudder shook its entire body. The hook was removed and the shark, seemingly lifeless, sank out of sight in the water. Thirty minutes later another strike was made with a baited hook. There was a short struggle and the shark was reeled in. It was the same shark. Blood was still oozing from the bullet wound.

When attacking humans they have been known to hang on with incredible tenacity.

TEETH

No living creature can match the shark for tooth-and-crunch power. Using a cylindrical 'bitemeter,' Dr Perry Gilbert estimates the shark's biting pressure to be in the vicinity of 18 tons per square inch.

Each species of shark has its own distinctive shape and size teeth. There are the cockscomb-like teeth of the tiger, the triangular teeth of the whaler, the prong-like teeth of the grey nurse. These have a sharpness of edge Gillette or Wilkinson might envy. But whereas the razor blade has a continuous edge, the shark's tooth, with its fine serrations, is more like a bread-knife and slices in the same fashion.

There is no need to waste sympathy on a shark that leaves several teeth behind after biting into the hull of a boat or losing several 'choppers' after chopping on a steel observation cage. Sharks' teeth are stepped in four or five — sometimes more — ready-for-action rows. If one or a number of teeth become worn, damaged or lost, the rows move up like overlapping steps in an escalator. This may happen within 24 hours.

The teeth action resembles that of a pair of scissors with the blades barely touching. There is none of the grinding or chewing that humans do. One Sydney doctor says the wounds resemble those of someone run down by train wheels.

Not all sharks bite in the same way. It depends on the shark and who or what it is biting. When it opens its jaws, the lower jawbone is thrust forward while the snout is drawn back and up until it makes almost a right angle with the axis of its body. One persistent fallacy is that the shark turns on its side to bite. Even Aristotle believed 'These fyshe turn on the back in order to take their food.'

Shark teeth are surrounded by a hard enamel. Because of this records are rich with examples of fossilised shark teeth. The teeth from fossil shark were used in Elizabethan England as poison gauges. Such was the widespread fear of poison that Renaissance princes had a fossil tooth mounted on a gold base which became part of the tableware. The method of use was simply to dunk the tooth into the food or drink first. If poison was present the tooth changed colour and someone was for the gallows.

SKIN

A shark's teeth are its front-line weapons. It can also cut with its sharp fin, thrash with that powerful tail and strike with force with its body. This all-round marine missile also has a protective armour around its body ten times rougher than the roughest sandpaper.

Most fish have scales. Shark skin is a mass of hard enamel-coated mini-teeth which under a microscope vary in shape with the species. They can give a swimmer a nasty graze that might well draw blood. These rasp-like denticles are useful in defence. They will often burr the sharpest knife, and harpoons, fired at close range, have been known to bounce off.

The Roman soldiers' helmet, called a galea in Latin, is closely related to the Greek word gateos meaning shark. Shark skin frequently formed part of the helmet because of its toughness. Shark leather has been used in various ways for centuries. It makes a joke of those adventure movies showing a diver armed with a knife in tropic waters, ripping open a killer shark's belly.

Thor Heyerdahl wrote, 'Our respect for the shark increased when we saw the gaffs bend like spaghetti when we struck them against the sandpaper armour on the shark's back.' Dr Coppleson, in his book *Shark Attack*, quotes an incident of shark attack described in the *Medical Journal of Australia*, 31 July 1920. So little was known about sharks at that time that what was believed to have been teeth marks on the victim's body, was in fact, rasps from the shark's hide.

'A.B., aged 20, and six other men set out on 8 March 1920 in a yacht in Cleveland Bay to take stores from Townsville to the Bay Rock Lighthouse. They made a fishing trip at the same time. When about halfway between Bay Rock and Magnetic Island, the boat overturned and sank. The occupants decided to swim to Bay Rock, a distance of about 3 miles. A.B. had been knocked on the head with a mast and "went out to it". After coming to, he started to swim. He had been swimming for about 30 minutes, when he was attacked and bitten by a shark. The shark evidently could not get sufficient grip in the region where it set its teeth. It tossed A.B. out of the water a distance of several feet. The shark did not trouble him again. He was then able to continue his swim and eventually reach Bay Rock. Three of the seven men were lost, but the others reached the rock without being molested. Cleveland Bay is infested with sharks.'

If the shark had indeed seized the man in its jaws and tossed him some distance out of the water, the man would surely have lost a lump of flesh. The wounds were superficial and Coppleson believed they were in reality, the results of contact with the shark's skin.

CHAPTER FIVE

▲▲▲▲▲▲▲▲▲▲▲▲▲

Shark habits

In the 1880s there were 160 known species of shark. Over a century later we have identified about 200 more. Today's sophisticated diving gear, which grew out of the needs of World War II and Jacques Cousteau's invention of the Aqualung, has brought man and shark face to face in greater numbers than ever before.

Skindiving, which increases in popularity each year, is becoming a major pastime. We will have to come to terms with this strange and fearsome creature. The more we understand it the easier it will be to do so. Already experienced divers have come to accept the shark in the way most of us accept the motor car when crossing the road. We remain watchful and know that if we don't get out of the way we are likely to get injured.

Underwater encounters are usually polite. The shark casts the merest glance at the skindiver as it cruises serenely past. But the diver knows there is always that tiny possibility that some idea will pop into the shark's brain that will cause it to act differently. The experienced underwater fisherman has never allowed his familiarity with sharks to let him treat them with offhand contempt. He knows only too well the shark can never be completely trusted.

There is a widespread belief that sharks are insatiable gluttons. This is because of the weird and wonderful assortment of objects found in a shark's stomach and the greedy way they are known to swallow bait at one gulp. In captivity and in simulated natural conditions a shark will eat, on average, an amount of food equal to 3-14% of its body weight — per week.

The shark is on the highest rung of the oceanic life-ladder. Each moving speck of life is devoured by a fish larger than itself right up to the supreme predator, the shark. They kill for survival. They do not kill for sport or gain or for sheer blood lust. At Sao Miguel in the Azores in 1964 a 2.4 metre shark was found in the stomach of a 16 metre sperm whale.

The shark is ceaselessly on the prowl. This is due partly to their lack of a swim bladder — they must either swim or sink — and partly to the necessity to keep an eye out for the next meal. From birth it has learned to be in a constant state of readiness; its whole life is ambush, chase, attack.

It does not follow that the shark is constantly hungry. When first taken into captivity it sometimes takes endless coaxing to get them to start eating at all. As big game fishermen know only too well, the shark's behaviour when faced with food is entirely unpredictable. A hungry shark will bypass a particularly tempting bait. Another shark who has just gorged itself, will seize it as though it hadn't eaten for a month. It is of little use trying to equate shark behaviour with the human drive. Our appetite and hunger patterns are totally different to the sharks' feeding pattern, which is still not entirely clear to us.

Sharks are not fussy when they eat; the time of day or the state of the tide. Neither are they particularly fussy about what they eat. Their diet appears to consist of just about any living creature found in the sea including sharks of their own species. Shark cannibalism is aroused when they find one of their less fortunate mates entangled in the nets. Nevertheless it is said that various species of shark do have their particular favourite morsel. For instance the hammerhead has a taste for crayfish, crabs and other crustaceans — when it can find them of course.

It is difficult to ascertain the relative importance of sight and smell in relation to a shark's appetite. Smell certainly enables them to discriminate between food and non-food objects although all caution is thrown to the wind when the shark is engaged in a notorious 'feeding frenzy.' The phrase was aptly coined by US marine biologist Professor Perry Gilbert. It describes the sudden release of collective ferocity, a delirium of destruction like humans running amok or going beserk. In this lunatic condition sharks will attack with indiscriminate fury whatever comes their way: objects, wounded whales, humans.

Gilbert describes such a scene in *Scientific American*, July 1962.

'The sharks first slowly circle at a distance of 2 to 3 metres. Then, as they swim faster, the circle tightens and presently one shark moves in for the first bite. Contrary to popular belief, the shark seldom rolls on its side. Braking its forward motion with its large pectoral fins, the shark points upward slightly as its mouth makes contact with the bait. It opens its jaws wide, the lower jaw dropping downward and the upper jaw protruding markedly from beneath the thin upper lip. If the first bite does not give the shark adequate hold, it bites a second and third time until it anchors its teeth deeply. Then it closes its jaws and shakes the entire forward part of its body violently from side to side until it has torn 4 to 7 kilograms of tissue from the marlin. Still shaking its head vigorously from side to side the shark swims away.

'As the blood and body juices of the marlin flow from the wound, the other sharks become more and more agitated and move in rapidly for their share of the meal. Three or four sharks will attack the marlin simultaneously. A wild scene, sometimes called a "feeding frenzy" now ensues.'

Substitute marlin for wooden boxes or anything you like, the biting and the tearing will be equally frantic. The attack is likely to end just as abruptly and the sharks will return to their normal behaviour.

The shark is often accompanied on its travels by small fish which are seen swimming in perfect V-formation in the bow wave of the shark. These are 'pilot fish'. It is widely believed they lead their protector to prey. This is unlikely. Sharks are capable of locating their own prey. But the pilot fish and the shark have a nice relationship to their mutual advantage. The pilot fish find plenty of food in the shark's leftovers from its meal. They are also protected from larger fish who would surely move in and devour them if it weren't for the presence of the shark. In return they pay their way by ridding the shark of clinging parasites and even go so far as to clean the monster's jaws after it has eaten.

Another fish that accompanies the shark is the small remora or sucking fish that fastens on to the shark and goes along for the ride like a hitchhiker. They can only do good for the shark by eating the small crustacean parasites that bury themselves into the gills and skin. The remoras too feed on morsels of food left over by the shark.

Although the shark may be constantly in search of food it does not mean it starts eating when it finds some. One fascinating aspect of undersea telepathy is that smaller fish sense when the prowler is not hungry and even make a corridor in their shoal to let him pass through. It is reminiscent of such land creatures as the zebra who senses the nearby lion is not hungry and continues to graze peacefully.

Sharks will digest their food rapidly — in one quick gulp — or slowly, depending upon various conditions. A tiger shark who died on 23 September 1950, one month after being taken into captivity at Taronga Park Zoo in Sydney, was found to have two 1.2 metre dolphins, in a perfect state of preservation, inside its stomach. Sir Edward Hallstrom entered in the log, under cause of death, 'Dolphinitis'.

In a number of instances human limbs have been found in the stomach of sharks caught in Australian waters. It has helped identify the shark as the attacker of a particular individual. In the case of Mary Passaris, 22, attacked off Broome in Western Australia on 16 May 1949, her arm, with a telltale ring on the finger, was found undigested in a shark caught five days later.

It was a scar that identified the arm of Norman Girvan, one of two young men taken at Coolangatta in 1937, when the shark was opened up. A tattoo helped to identify the owner of the arm inside the tiger shark in the Coogee Aquarium in 1935. No book about attacks in Australian waters would be complete without details of the world-famous Shark Arm Murder. Even today's most avid TV crime fan would find the story hard to swallow. The series of coincidences that opened up police investigation of this celebrated case went beyond the most bizarre fiction.

THE COOGEE SHARK ARM MURDER

On the morning of 18 April 1935 a 14.2 metre tiger shark ceased struggling at the end of a line, a kilometre off Coogee Beach. The bait had been left overnight by the Hobson brothers who operated the Coogee Aquarium. A shark, less than half the size of the tiger, had apparently taken the bait in the first place. Its vibrations were picked up by the tiger shark who raced in for a cannibalistic feed. In no time all that was left of the smaller shark was its head. In the process the tiger got tangled up too.

Pleased with the unexpected catch the Hobsons conveyed their prize to their aquarium. But their new star performer failed to live up to expectation. It wasn't in the best of health and moved slowly, listlessly, around the pool. Oxygen was pumped into the water and the shark appeared to perk up — just in time for the Anzac Day sightseers. At 4.30 on Anzac Day afternoon, Thursday 25 April 1935, a dozen spectators saw the slow-moving shark suddenly spurt forward, make three rapid turns and leave behind a foul-smelling scum that floated on top of the water. The shark had 'thrown up.' The distasteful look on the faces of the onlookers turned to shocked disbelief when someone shouted, 'Hey look! It's somebody's arm!'

Police officers from Randwick Police Station arrived. The human arm, with its length of rope tied to the wrist, was gingerly removed from the pool. The arm had a tattoo of a pair of sparring boxers. The arm was conveyed to the city morgue and the following day Dr Victor Coppleson, the leading authority on shark wounds, accompanied the government medical officer for an investigation. It was quickly ascertained that the arm had not been bitten from the body by a shark. There were no jagged edges to the flesh wound. The arm had been removed with a knife — by an amateur.

Next day the tiger shark, which was dying anyway, was destroyed. It could have turned out a real money spinner after having disgorged a human arm. What was even more incredible, shark experts decided the arm had in fact been lodged in the stomach of the first shark; the smaller one.

The arm was so well preserved that fingerprint experts from the CIB were able to link it with James Smith, 40, one-time SP bookie, billiard hall operator and bankrupt builder with minor convictions. CIB detectives began putting the pieces together. Smith had been recently involved in business dealings with popular North Sydney boatbuilder, Reggie Holmes. The pair had also been involved with a high-priced yacht, *Pathfinder*, which had sunk under highly suspicious circumstances. Claim for insurance money was made then abruptly cancelled. On 7 April, 11 days before the tiger shark was caught, Jim Smith had left his home in Gladesville, telling his wife he was going fishing. He was never seen again.

The focus was now on a third man — Patrick Brady. Brady was an expert forger with a string of convictions going back 20 years. He was last seen in the Cronulla area. The barman at the 'Cecil' told police Brady was seen drinking with a man answering Smith's description. Brady had rented a beachside cottage named 'Cored Joy' for 30 shillings a week. The estate agent said the house was left spotless but there were a few curious changes. An old tin storage trunk was replaced by a smaller one. A mattress had also been replaced. An anchor was missing from the boatshed and so were two heavy window weights.

The search was on. Police trudged across sandhills, crisscrossed the waters of Port Hacking and even tried aerial reconnaissance for the tin trunk that surely carried the remains of James Smith. Local cabbies reported driving Brady to several locations in the city. One driver had taken him to an address at McMahon's Point: the home of Reginald Holmes.

On the evening of 16 May, exactly three weeks after the arm was seen floating in the Aquarium, two police cars pulled up outside a block of flats in North Sydney. Mrs Brady opened the door. Her husband was in the bedroom. Under intensive questioning, the tough, quietly-spoken Brady confirmed all the facts the police already knew. He had no idea what had happened to Jim Smith after he saw him at the cottage on 9 April. Brady also said he knew the boatbuilder, Reg Holmes. Holmes in turn, emphatically denied knowing Brady. He persisted even when the two were brought face to face.

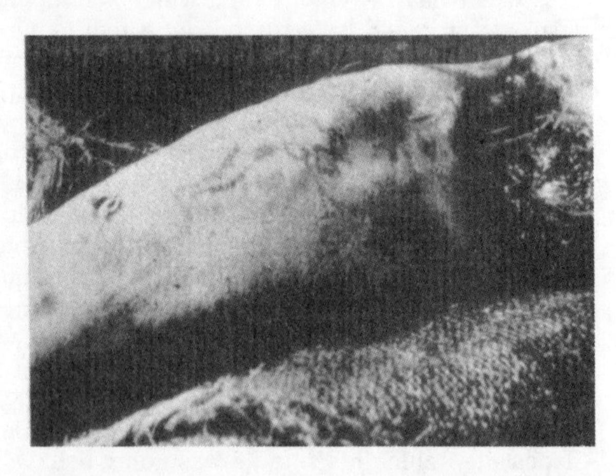

James Smith's disgorged arm

Patrick Brady was formally charged with the murder of James Smith. But Mrs Grace Brady was having none of it. She was convinced of her husband's innocence and believed he was covering for Reginald Holmes. She told as much to the police. Brady found himself obliged to make a further statement. This time he said that Smith had left the cottage for the last time in company with Reginald Holmes and another man. Once more the police wished to question the 34-year-old boatbuilder. But Reginald Holmes had already made other plans.

In the chill autumn dawn he launched a newly-built speedboat and steered it through the rippling waters of Sydney Harbour. Slowly he raised a .32 calibre automatic revolver to his head. Then he pressed the trigger. The impact threw his body over the side. He reappeared a moment later grasping the side of the boat. Blood streamed from a head wound. In his faltering gesture the bullet failed to penetrate his skull and instead had cut his temple. A strange report came through to the water police that a speedboat was racing around the Harbour driven by a man with blood all over his face. What followed was a hair-raising chase in the best traditions of the perennial police drama, only this one was on water. Two powerful police launches came in hot on the tail of the speedboat which the wounded man handled with great skill. It led them on a breathtaking course doubling in and out of bays, weaving around boats and buoys, careering around jetties. It lasted two whole hours. Finally Holmes cut the motor near Watsons Bay and called out to the police launch *Nemesis* for bandages. He claimed he thought it was someone other than the police chasing him. He smelt strongly of alcohol.

The sick tiger shark in the Coogee Aquarium just before it disgorged the arm of James Smith.

Holmes spent four days in Sydney Hospital under police guard. Almost completely recovered he made several statements to the CIB. In doing so he well and truly 'dobbed' Brady, who he said, had told him he had got rid of Smith's body. 'I dumped it in a tin trunk outside Port Hacking.' He also told police Brady had told him to keep his mouth shut, otherwise 'If I'm not able to get you,' Brady threatened, 'one of me cobbers will.'

In telling all this to the police, it looked as though Reggie

Holmes was really sticking his neck out. Holmes was allowed to go home. The police had got themselves a star witness; yet surprisingly they did nothing whatsoever to protect him.

The inquest was set down for 12 June. Shortly after 1 am on that day a police patrol car drew up beside a Nash sedan parked at Dawes Point, just below the three-year-old Harbour Bridge. The front door was open, the headlights blazing. They thought the man slumped over the wheel was drunk. He was dead. He had been shot twice in the left side. The man was Reginald Holmes. Patrick Brady was already under lock and key. Could it be that one of those 'cobbers' of his had made good the threat? Two men were later arrested. Both were acquitted. So too, through lack of evidence, was Patrick Brady.

The body of James Smith was never found. Apart from the anonymous telephone calls, the threatening letters and the whispers surrounding prominent Sydney people that followed in the wake of the 'Shark Arm Case' for years after, the affair achieved little beyond a place in the world annals of unsolved mysteries. The Nemesis that first appeared in the shape of a sick tiger shark had come to nothing — or had it ...?

THE ROGUE SHARK

Fortunately sharks are just as likely to bump people as bite them. Scores of surfers have reported being nudged, nicked or knocked off their boards by a bump from a shark's snout. Innumerable times defenceless board riders have been thrown into the water this way, yet, although the situation is ideal for an attack, there appears to be no report of a shark biting a 'bumped off' surf rider. One explanation is the shark is just being 'nosey'. It has even been suggested the shark is having a little game. After all, if dolphins can play around why should a shark be such a miserable old predator! The difference is of course that the dolphin has a much greater intelligence.

In January 1949, lifesaver Don Dixon was out on his surf ski waiting to catch a wave. Next moment there was a splash and a thump and Don was in the water. In desperation Don kept scrambling on to his ski and slipping off again. He did this all the way back to Mona Vale beach with the shark eyeing his performance as it escorted him part of the way to the shallows.

Sharks are as cautious as they are inquisitive. They will endlessly circle a puzzling object, sometimes moving in to nudge it. A smart smack on the snout with a paddle has made many a shark turn tail and take off. In other circumstances this has understandably infuriated the shark.

In the *Medical Journal of Australia* 15 April 1933, the distinguished Sydney surgeon, Dr Victor Coppleson (he was knighted before he died) wrote the world's first major thesis on shark attacks. In the first comprehensive book on the subject, *Shark Attack*, published by Angus and Robertson 15 years later, Dr Coppleson propounded the theory of the 'rogue' shark. The pattern is not unlike those stories about the Bengal tiger who, having once tasted human flesh, turns into a maneater and prowls remote Indian villages at night in search of a meal. The shark does not have the cunning of the tiger. It does not lie in waiting. It does not stalk its prey in the same fashion. The lone shark that has developed a taste for humans patrols an area on the lookout for an opportunity to savage other humans.

Dr Coppleson distinguished the 'local rogue' from the long-range cruising shark. It has already been ascertained there is a principal population that lives and breeds in the area and the accessory population, outsiders who have permanently lost their way and have come in from another geographic zone. The rogue appears to come from among the outsiders.

Dr Coppleson cited many instances of 'grouped attacks' — several of which are reported elsewhere in this book — both in Australian waters and elsewhere. He highlights the Coogee-Bronte attacks: four in the space of three years in the early 1920s. In roughly the same period, 1922 to 1925, in the West Indies, there were five attacks on the beaches of San Juan, Puerto Rico, in which six people were injured, two of them fatally. There had never been an attack before or since at Bondi, yet in less than ten months between 1928 and 1929 there were three, two of them fatal.

Only four days after 16-year-old Thomas MacDonald was attacked at Byron Bay in northern NSW in October 1937, two men died in a terrible attack at Coolangatta, 50 kilometres north.

At Cartagena, on the Caribbean coast of the south American republic of Colombia, there was a series of shark attacks in August 1951. Four people were killed within a week and several others injured.

Then there was the episode of the 'mad shark' of New Jersey that killed four people and injured a fifth in the space of ten days and had America up in arms in July 1916. Coppleson believed the rogue's patrol generally to be a dozen miles from the spot of the first attack. The danger period for humans could last up to two years.

More recently, in June 1993, all 30 of Hong Kong's public beaches were closed after two fatal attacks on swimmers in relatively shallow water at a popular swimming beach in Sai Kung within 12 days. Post mortem examinations indicated both attacks were consistent with the bite and attack pattern of a tiger shark up to 7 metres long. Two years previously a similar fatal attack in the same area had been the first recorded in 11 years. The deaths prompted warnings from environmentalists that pollution and over-fishing in the seas around Hong Kong were making sharks hungry and desperate — and therefore liable to set up a feeding pattern off Hong Kong's beaches. There is, however, a body of experts who refute the theory of the rogue shark. Their main argument is that the shark just hasn't the intelligence to hang about like that!

ATTACK THEORIES

The theory that sharks attack whites in preference to dark-skinned people was first suggested in a book published in 1868 called *The Ocean World*, by a well-known, but perhaps best forgotten, naturalist of the day. When given a choice, he claimed, sharks 'prefer Europeans to Asiatics and both to the Negro.'

This piece of undersea racism is hardly born out by the tragic stories of the slave trade, when black Africans were piteously pitched over the side at the merest whim, and sportive sailors dangled black bodies from the yard-arm to see how high a shark would leap to get at them.

It is said that Englishmen dwelling in the tropic climes of the far-flung Empire in the early 19th century surrounded themselves with coloured servants when they went in for a dip, convinced their servants would act as a protective shield against a charging shark.

The records of native divers fatally attacked are not kept as meticulously as in more populous regions. Any real advantage is likely to be psychological. Native divers are wary of sharks. They are mentally adjusted to the possibility of attack. Should the worst eventuate the shock to their system is not as violent as it is to the average bather.

Many old hands believe that a fully-clothed person — as long as his clothing isn't white — washed overboard is less liable to be set upon by a shark than a bather in a pair of trunks. More likely clothing would prevent severe scuffing of the skin should a particularly chummy shark decide to get up too close.

Finally there is a theory which is not so easy to test, even when the shark is in captivity. Will it attack a male human in preference to a female? Even allowing for the fact that generally men outnumber women in the ocean, statistics show that relatively sharks show a marked preference for males. One can only draw one's own conclusions.

MATING

Sharks are divided into roughly three groups according to the way their embryos develop. These are:

(i) Oviparous (dropping eggs that hatch outside the body)
(ii) Ovoviviparous (bearing live young from eggs hatched within the body)
(iii) Viviparous (bearing live young)

Most shark species are ovoviviparous. In all species, fertilisation is internal. The male of each species has distinctive claspers which match the opening in the female of that species only. They cannot enter the female of any other species. During copulation both the male claspers or one at a time are inserted.

An eyewitness account of two captive sharks in copula is described as follows.

'The sharks were side by side, heads slightly apart but the posterior half of their bodies in such close contact and the swimming movements so perfectly synchronised that they gave the appearance of a single individual with two heads, as they swam in slow counter-clockwise circles around the pen ...

The sharks were together when their observers left half an hour later.

Mating behaviour appears to differ with the species of shark, involving more violent activity and even 'love bites'. The young shark, or pup, is prowling and snapping almost from birth.

An unusual characteristic of one species of shark was discovered by a scientist while examining a pregnant grey nurse shark. An unborn baby bit him on the hand. He subsequently found that the first baby to emerge from its egg within the uterus feeds on the other babies as they hatch. The grey nurse female has two separate uteruses: consequently only two young sharks are born. There are several species of shark in which intra-uterine cannibalism occurs but there are no other known cases in the animal kingdom.

CHAPTER SIX

▲▲▲▲▲▲▲▲▲▲▲▲▲

Scenario for attack

Should any insane person really want to be attacked by a shark, the simplest approach would be to dive straight off the side of a boat into the depths of Middle Harbour or plunge into the centre of the Georges River. Of course there is always Lane Cove River or perhaps Pittwater might be handier. You won't find it quite so easy along any of the Sydney metropolitan beaches. These have been mesh-protected since 1937, but there are ocean beaches beyond the eastern seaboard cities which are not protected.

If you really would prefer to meet your fate in the surf, however, it would pay to choose the ideal conditions for shark attack. January is by far your best bet. December could be your next choice, followed by February and March which are about on a par. The kind of day you select should be hot, preferably humid. Lots of cloud about. A little rain wouldn't do any harm. The warmer the water the better. Anything from 20oC and up will do. Depth is left to you; from knee-deep to open sea. Still, it might pay to linger around the metre mark. Hopefully this will be anywhere between 10 to 20 metres offshore. It matters little whether the sea is rough or calm.

Go alone if possible. If there are other bathers around keep well away from them. If a group of mates catch up with you and insist on joining you, suggest you all try some body surfing, only make sure they catch a nice big wave and you are left alone out there treading water.

Your chances of being taken are much greater if you are male rather than female; about 22 to one. And it is no good looking around hopefully for your attacker. Almost certainly the shark will be upon you before you even have an inkling of its presence.

You will have to take pot luck on the type of shark and its size. But be assured it will be in excess of 1.5 metres and whatever the species, the papers will probably say it was a grey nurse. Ironically, it is precisely because you want to get yourself attacked by a shark, that the chances are you won't be. For in some mystical way it is your very confidence — the confidence of an experienced skindiver — and your lack of fear in coming face to face with a killer that will probably discourage it from coming face to face with you.

A shark senses fear. It senses helplessness: a wounded fish, a drowning man. But don't count on it. Never count on anything connected with a shark. Each shark assault is a highly individual affair. Broad generalisations just aren't good enough. There are too many factors involved. These concern not only time and place and conditions but also a variety of factors relating to the shark itself: species, sex, state of health, whether it is hungry. Even then its actions will be unpredictable. The more one studies shark behaviour the less one seems to know about them.

The chance of attack is minute. Perhaps because of this you can drive to any of a number of popular picnic spots along the shark-infested Georges River on a hot Sunday, to see scores of adults and children splashing a few dozen metres from a sign which reads DANGER SHARKS. Yet, when a fatal attack hits the newspaper headlines, it holds all the fascination of a gruesome murder.

IN DEEP

One persistent fallacy relating to shark attacks is water depth. 'Don't go out too deep!' is the constant catchcry of parents. Thousands are convinced that provided they do not venture further than waist deep, they are as safe as houses.

Perhaps the most frightening aspect of all is the shallowness of the water in which sharks have been known to attack. On 17 April 1949 Richard Maguire, 13, was attacked at Ellis Beach 30 kilometres north of Cairns. The water was 0.9 metres deep. He was dead when they got him ashore. More incredibly, the shark, which was later caught, was 3.3 metres long.

The shark that attacked another youngster, 10-year-old Douglas Lawton in the same depth at Sarasota, Florida on 27 July 1958 was less than half that length. The incident has similar grisly overtones to the attack on Ray Short at Coledale on the south coast in 1966. It clung on tenaciously even after the boy was brought ashore and had to be knocked out with a rock before it would unlock its teeth.

SAFETY IN NUMBERS?

Answer: None. The difference is that the shark who decides to attack has more to choose from. The odds are more in one's favour. Crowds for that reason are safer than swimming out there alone. The noise and din — and what a din — of a crowd should surely scare off the hungriest of sharks. It didn't scare off the shark that fatally mauled surfing champion Ray Land during the noise and commotion of a surf carnival at Bar Beach, Newcastle in January 1949.

Although a shark usually attacks the surfer on the edge of the crowd, they have also torpedoed in on one particular person right in the very centre of a crowd of bathers. Why that person? There have been theories ranging from scent (even the brand of suntan lotion!) to colouring, to movement in the water. If the precise truth were known, it would be a major step towards finding the Golden Fleece of all shark biologists — the perfect repellent.

DISTANT SHORE

A US Shark Research Panel investigating the activities of 360 victims minutes before being attacked, found 20% had been wading in water anywhere from knee-deep to chin-high, while 57% were actually swimming.

During the investigation by the same panel, one unexpected fact surfaced. Almost 40% of the attacks occurred within 30 metres from the beach. The further out the smaller the number. At 90 metres to 400 metres the percentage was less than half of the attacks within that first 30 metres. The figures related to the numbers of people swimming at these distances appeared to indicate that the further one swam from the shore the safer. This in no way tied-in with the lesson of attacks on the one swimmer. Yet another shark ploy to bamboozle humans!

WATER TEMPERATURE

Dr Coppleson firmly believed in the relation of water temperature to shark attack. In almost every instance of attack around Australia where it was possible to report water temperature, the attacks occurred in temperatures of 21°C and higher.

Elsewhere, attacks have taken place in waters of lower temperatures. Since the late 1960s four of the victims of attacks in California were in water around the 13°C mark. Water temperature undoubtedly is a significant factor in shark attacks. Warm water seems to heighten the metabolism of cold-blooded creatures. Cold water calms them. In Australian waters at least, the colder the water the less likelihood of attack.

Forty years ago (a drop in the ocean of shark evolution) a Sydney shark aerial observer noted that the patrol log recorded a complete absence of sharks when the winds were blowing from a southerly direction.

It was also noted that there were more sharks spotted north of the Harbour than south. The enormous growth of population along the northern beaches in that time might have affected this unwanted popularity.

TIME TO ATTACK

Thor Heyerdahl observed from the Kon Tiki that the sharks following the raft became more aggressive towards dusk. Certainly attacks in Australian waters have occurred later in the day. There appears to be a sharp increase from about 3 pm onwards with the peak hour between 3.45 and 4.45 pm. It is difficult to ascertain whether it is safer to swim at night than in daylight. So few people do, and wisely so.

Ben Cropp has written that in encounters with thousands of sharks he has never seen one take a fish swimming by during the day. This does not apply to a wounded fish. The killer instinct in the shark finds the weakened fish irresistible at any hour. When sharks have their proper meal they generally do so at night, filling their bellies two or three times a week.

TASTE FOR HUMANS

If sharks really wanted to gobble up humans none of us would dare surf, dive or even go fishing. Fortunately sharks do not appear to find human flesh particularly tasty. Nor do they seem to acquire an enduring taste. There are many instances when skin-divers have been attacked because they carried dead or dying fish around their waist or at the end of a spear, or in some instances because they have handled fish inside the boat and their hands carry the smell. It is also true a shark may attack in self-defence or because it is offended in some way. What is not true is that the shark prowls in search of human flesh. Even the rogue shark, who has already sampled some, does not hang around strictly for reasons of appetite.

One curious factor of shark attack is that once the shark has taken a piece of its victim, an arm or a leg, it generally appears satisfied. Not often does it return for more. Nevertheless attacks recorded in this book show it sometimes does. This adds to the widespread belief that the shark, suddenly surprised, might snap out instinctively. Attracted by an unexpected vibration it seizes that arm or leg beckoning from the silvery ceiling of the water's surface. It certainly doesn't say to itself, 'Beauty! There goes my favourite meal!'

PROCESS OF ATTACK

Like the car that slams you, it's the shark you didn't see that is likely to attack. For this reason experienced divers emphasise the importance of awareness of what is going on around you. The attack frequently comes from below and behind. It may begin with a nip on the calf or thigh. Having 'tested' its prey, encouraged by the flow of blood and stirred by its victim's wild panic, the shark circles below and charges in for the second and third time, its ferocity increasing, taking bigger bites each time.

An attack can quite easily bring several other sharks rushing to the scene in this highly competitive nether world. This is perhaps inevitable. The greatest attention-getter for a shark is a struggling fish. Underwater fishermen say they might swim for hours without sighting a shark. The moment a fish is speared a shark will appear as if by magic.

CAUSES

Shark behaviour has sometimes been likened to that of dogs. In fact 'dog fish' was the term for shark originally used by the Romans. You walk down a street; a dog sits outside his gate. His size, his appearance give no indication of how he will react when you pass. He might ignore you. He might follow you, chummily or menacingly. He might bark himself hoarse. Rarely, but there is always the possibility, he will snap. Even then it could be bluff. Just that one time he will bite for real.

His reaction will depend on several factors, whether you make a sudden movement, whether you falter or show fear. It might be something you are carrying or the way you swing your arm. It might even be the way you call to him or whether you give him a wide berth or pass by nonchalantly. You never know. The cute, harmless-looking dog might turn out to be more aggressive than the big, fierce-looking fellow.

In his book, *Lord of the Sharks*, Franco Prosperi states that where people are rarely seen by sharks, they are more hesitant before they attack. He believes cause factors differ with the species of shark. Do they do so with the breed of dog?

Continuing with the canine analogy, that same dog is just as likely to act in a totally different manner when the next person comes along. Is it an individual's looks? Has it to do with one's aroma, pleasant to other people of course, but disagreeable to that dog? Does it go beyond these factors? Is there a touch of the supernatural somewhere, something in your particular aura that brings out the aggression in a particular dog — or in a particular shark?

But what about the instance in which the shark plucks its victim from among a crowd of bathers? This is somewhat different. It may merely be a matter of focus. Why does the ravenous lion pounce on a particular antelope in a fleeing herd? It can only take one at a time. Was it influenced by the antelope's size, the way it ran, its distance from the remainder of the herd?

Of ten cars racing over the speed limit along the express-way, why, why did the highway patrol have to pick on you?

CHAPTER SEVEN

Down in the drink

The perils facing survivors shot down or shipwrecked in shark-infested seas proved an unexpected hazard in World War II. It was an issue wartime governments found expedient to hush-up. Even the survival manuals played down the risks.

Land animals such as crocodiles, snakes and jungle cats were barely considered in escape plans. Sharks, however, have proved a powerful deterrent for potential escapees from more than one penitentiary, including Devil's Island, 11 kilometres off the coast of French Guiana; Alcatraz in San Francisco Bay; Pinchgut in Sydney Harbour; Rottnest Island in the Indian Ocean off Perth.

Allied servicemen, who, as boys had read blood curdling adventure yarns about killer sharks, found themselves in remote parts of the world faced with the very real horror of shark attack. It was the Americans who realised, like it or not, the facts had to be spelt out to personnel in tropic zones. Official information about sharks was not entirely comprehensive. Knowledge of shark behaviour was limited in the early 1940s. Added to this it was considered advisable to play down certain harsh facts. No sense in scaring the pants off the boys. A booklet entitled Shark Sense was written in a whimsical style accompanied by amusing drawings. In retrospect Dr Coppleson believed it was unwise to 'regard the killer shark as something that comes out of legend rather than actuality.'

One of the most hair-raising stories of individual survival took place during World War II when a military aircraft belonging to the neutral Central American republic of Equador, crashed over the Pacific in June 1941. The young pilot officer was in the sea 31 hours. Both his companions were injured. Five hours following the smash the Colonel died.

'Afterwards putting the corpse, which floated perfectly, in front of me I continued pushing it, with the objective of taking it out ... if we managed to reach land. When I had pushed the corpse of my colonel ahead of me in order to swim to it again, a strange force dragged the body and I did not again see it in spite of searching a long time among the waves. Sub-Lieutenant D., who was still living, and who had hold of me, made me reflect that it was foolish to wait longer and I continued towards where I believed the coast to be. Sub-Lieutenant D. lived perhaps four or five hours more until at the end, after having had moments before his death of a state of despair which is very painful to narrate, he died.

'Taking the same attitude as toward that of the corpse of my Colonel, I put the body of my companion in front of me and continued pushing him but not as far ahead as I had done previously with the other corpse. As it was a moonlight night and during some moments very clear, I was able to observe that strange figures crossed very close to us until at a given moment I felt that they were trying to take away the corpse pulling it by the feet, on account of which I clutched desperately the body of my companion and together with him we slid until the tension disappeared ... Once refloated, with despair I touched his legs and became aware that a part of them was lacking ... and continued swimming with the now mutilated corpse until the attack was repeated two times more and then, terrorized at feeling the contact of fish against my body, turned loose the corpse ... convinced that I would be the next victim ... As soon as it was light I could see the coast at a great distance but I had no hopes of reaching it because with the light of day I could see that various sharks were following me. When I moved my legs slowly, with the object of resting, I touched with my feet the bodies of these animals which were constantly below mine in order to attack me. I would then thrash the water and thus for a few moments the danger would pass. I continued swimming all day Friday until at sundown I found myself some 400 or 500 metres from the rock on the coast, and as I was already tired ... because of the undertow which existed I could not reach the rocks until after making a superhuman effort ... at which hour I do not know ...'

America entered the war six months after the Equadorian officer's harrowing experience. His account seemed to portend situations in which US servicemen might find themselves. This certainly turned out to be true. In the meantime, experiments began to find a suitable shark repellent. Various case histories were examined, including Dr Coppleson's 1933 thesis on 42 Australian attacks. Eventually scientists were to come up with a mixture to which they gave the lightweight name of Shark Chaser. If nothing else, the name sounded easy and effective.

Several years after the end of World War II, Dr G. A. Llano of Washington's National Science Foundation, made a study of almost 2,500 personal accounts of wartime flyers brought down over shark-infested waters. All things considered, the record was not too bad. There were only 38 shark sightings. In only 12 of these, did the shark attacks result in death or injury. There were undoubtedly others who never lived to tell the tale. The figures applied to airmen only.

Two other interesting facts emerged from the analysis. In 11 cases the sharks appeared within half an hour of the survivors landing in the water. In the remaining 27 instances, the sharks put in an appearance anywhere up to 24 hours later. The second fact was that no one mentioned the Shark Chaser which most of them carried in their survival kits.

The sharks usually kept their distance from the life-rafts, content to cruise in the vicinity. Only when there was blood or vomit in or around the raft did the sharks get excited and 'make passes' at it.

One group of survivors reported that for 17 agonising days the sharks never left them, growing increasingly bolder, even leaping out of the water and spraying the occupants or thumping the soft deck fabric with their tails. Later one afternoon, one of the sharks, a little over a metre long, actually slithered into the raft and took a bite out of one of the men, before they managed to fling it out. The man became delirious and died four hours later.

Some survivors began to realise, with relief, that sharks who frequently dived beneath their rafts were not trying to tip them, but were after the small fish who seemed to take refuge in its shadows.

One case history was experienced by a lone Navy pilot who spent eight days in a dinghy near New Britain. He was the only one to survive an all-out attack on his craft. The sharks showed up at sundown on his third day. A shark began striking the boat with its head. On its third approach the pilot shot it through the head at point blank range and it sank out of sight. The story was repeated on the fifth night. There must have been something about the craft or the man inside it that made them behave this way. On the evening of the sixth day a shark began bumping the dinghy with its nose. The man raised his .45 about 15 centimetres from its head but the gun would not fire. The slide had rusted.

'It then came to the surface again and made rushes at me, spinning the boat completely around 360 degrees several times.

'During all this I was holding grimly on and although I have no recollection of it, undoubtedly screaming! He came to the surface again, lay on his back, and began snapping at the boat. Never was I so grateful to mother nature for the placement of his mouth. I'd given up trying to load the .45 and was swinging it by the muzzle at him. I smashed him in the eye, on "his very vulnerable nose," and his "soft" belly. He turned over then and I started to pound him on the top of his head. He was as hard as steel there, and I later discovered I'd partly flattened the little steel eyelet on the butt of the gun where the lanyard is attached. He rolled over again still snapping at the boat and I remembered a capsule of chlorine so I tried to get it out of my pocket, with no success. I had dropped the gun and was searching my pockets with one hand and was hitting my voracious friend with my other fist. In that ill-advised action, I got two fair-sized splits in my forefinger, when one of his snaps and one of my blows became beautifully timed.

'Giving up the chlorine idea I seized the dye marker can and dumped some of it in his face, and the action ceased. The whole action had lasted from five to ten minutes. Whether the shark realised I was in the boat or whether he was merely venting his rage on this strange yellow object I will never know, but he made a savage and sustained attack and the ultimate end would have been the same, mine!

'After his departure I took stock of the situation and was depressed beyond all hope. The bottom of my boat had 18 holes in it. One was a slit about 4 inches long, another a round hole about the size of my fist. The rest were assorted shapes and sizes, all smaller. This was not so bad, but at the small end of the boat in the inflated part there were five slits all leaking badly. I must confess I gave up, believing I was doomed, drank all my remaining water thinking that at least I didn't need to be thirsty any longer.

'(The 7th evening) My gun worked okay and I killed or badly wounded two sharks and had no further trouble ... After my first encounter with sharks, I never even considered going over the side of the boat to avoid strafing ...'

Often the presence of the shark was a form of indirect torture. Dr Coppleson in his book Shark Attack, retells the story of the three British soldiers with scarcely any navigational experience who decided to risk the open sea instead of surrendering to the Japanese at Singapore in 1943. They set out in a 5 metre dinghy for a 2,500 kilometre voyage to Australia. They survived 125 days before reaching land. Sharks drove them to the edge of madness. Beaten down by the merciless sun they craved to cool off in the sea, but the sharks never left them, not for a moment. They cruised like vultures in their wake. The land they sighted after their terrible ordeal was not Australia. It was Sumatra, 200 kilometres from the spot from where they first took off.

There were several recorded instances when rescues were made in true Hollywood fashion, in which survivors were snatched from the very jaws of death. This happened to US Lieutenant Commander Kabat, who floated wearing nothing but a kapok life jacket, off Guadalcanal in 1944.

As dawn broke he felt a curious tickling sensation in his left foot. 'Slightly startled, I ... held it up. It was gushing blood ... I peered into the water ... not 10 feet away was the glistening brown back of a great fish ... swimming away. The real fear did not hit me until I saw him turn and head back toward me. He didn't rush ... but breaking the surface of the water came in steady direct line. I kicked and splashed tremendously, and this time he veered off me ... went off about 20 feet and swam back and forth. Then he turned ... and came from the same angle toward my left ... When he was almost upon me I thrashed out ... brought my fist down on his nose ... again and again. He went down about 2 feet swam off and waited. I discovered he had torn off a piece of my left hand. Then ... again at the same angle to my left ... I managed to hit him on the eyes, the nose. The flesh was torn from my left arm ... At intervals of ten or 15 minutes he would ease off from his slow swimming and bear directly toward me, coming in at my left. The big toe on my left foot was dangling. A piece of my right heel was gone. If he did not actually sink his teeth into me, his rough hide would scrape great pieces off my skin. The salt water staunched the flow of blood somewhat and I was not conscious of great pain.'

In the excitement of trying to attract the attention of a passing ship the naval officer forgot the shark. It struck again and took a lump from his thigh. At this moment he was seen. Several sailors began firing from the deck at the shark.

'A terrible fear of being shot to death in the water when rescue was so close swept over me. I screamed and pleaded and cried for them to stop. The shark was too close. They would hit me first.'

Two US navy flyers, Almond and Reading, had to ditch their plane in the central Pacific. Reading was knocked unconscious by the impact, but his radioman, Almond, eased him out of the cockpit and got him into his inflatable life jacket before the plane sank from sight. Lieutenant Reading gave his report.

'After I came to, A. told me the plane had sunk in two minutes and he didn't have time to salvage the life raft. He pulled both our 'dye markers' and had a parachute alongside of him. He did not have any pants on at all except for shorts ... We soon lost the chute and began drifting away from the dye. It was within a very short time (about 1/2 hour) when sharks were quite apparent swimming around us. A. and I were tied together by the dyemarker cords and it made it difficult to make any headway. An hour later we heard aircraft and I said to A., "Let's kick and splash around to see if we can't attract their attention". It failed, but suddenly A. said he felt something strike his right foot and that it hurt. I told him to get on my back and keep his right foot out of the water, but before he could, the sharks struck again and we were both jerked under the water for a second. I knew that we were in for it, as there were more than five sharks around and blood all around us. He showed me his leg and not only did he have bites all over his right leg, but his left thigh was badly mauled. He wasn't in any particular pain except every time they struck I knew it and felt the jerk. I finally grabbed my binoculars and started swinging them at the passing sharks. It was a matter of seconds when they struck again. We both went under and this time I found myself separated from A. I also was the recipient of a wallop across the cheekbone by one of the flaying tails of a shark. From that moment on I watched. A. bob about from the attacks. His head was under water and his body jerked as the sharks struck it. As I drifted away ... sharks continually swam about, and every now and then I could feel one with my foot. At midnight I sighted a YP boat and was rescued after calling for help.'

The stories of survivors of wrecked ships and aircraft in both war and peace who have died, not from drowning, but from shark attack, have been hushed up for obvious reasons.

The details of one appalling disaster in which the greatest number of people were killed by mass shark attack were released long after the end of World War II. The American cruiser, the USS

Indianapolis had been on a secret mission involving the delivery of the world's most awesome weapon of destruction — the atomic bomb. Some may see the ensuing tragedy as a kind of symbol.

The Indianapolis had completed its speed run from the USA to the tiny mid-Pacific island of Tinian. There it delivered the atomic bomb which was to be dropped on Hiroshima, 6 August 1945. Still under sealed orders she was now on her way to Leyte in the Philippines. Moments after midnight on 29 June the giant cruiser was sighted in the moonlight by a Japanese submarine. The two torpedoes that streaked in on the battleship were right on target. She began to sink almost immediately. The lifeboats had been lashed firmly for the speed run, there was no time to cut them loose. One thousand sailors threw on life jackets and plunged into the dark sea.

All that was left of the proud ship was bits of floating debris and approximately 900 of her complement of 1,200 men, bobbing about in the water, shouting wisecracks to each other to keep up their spirits. Senior ship's officers felt they had been lucky. But what a handful of those officers knew and the remainder of the crew did not know was that no distress message had been sent. Worse, it could be many days before the cruiser was listed as missing. So secret was the mission, the US Pacific Fleet was not even aware she was in the area.

Within the hour a dorsal was seen weaving among the shipwrecked men. Daybreak revealed several shark fins in the vicinity. The men, bobbing helplessly about in the sea, made jokes about them, but these had a hollow ring. In truth they were terrified. A number of the survivors were injured and had been bleeding steadily from their wounds. Already several were dead, their bodies hanging limp in the life jackets.

All through that first day more sharks appeared. The men decided their only defence was to form tight groups and try to scare them off if they came close. As the sun went down, the silhouette of scores of shark fins moved restlessly to and fro against the red sky. At first the tactic worked. The men nearest the shark would shout obscenities and thrash the water with their arms. The shark drew back. But now, when the creature had singled out its victim no amount of hullabaloo would keep it from its objective. There would be a bloodcurdling scream and the man would disappear in a red swirl. Even the most devout atheist prayed to God.

Next morning the number of survivors had been reduced by more than a hundred. The jokes had died too. Except for an occasional exchange, each man drifted in his own private hell. As the tropic sun beat down upon them, some began to go mad with thirst. A few began to drink the ocean water. Others followed. They died of convulsions soon after. Occasionally a man imagined he saw a distant island and would begin swimming towards it. The sharks moved in on him before he had got more than a few hundred metres from his group.

One of the survivors, Quartermaster Robert P. Gause, recalled that in the early morning light the man next to him was slumped over with part of his head in the water. 'I thought he had gone to sleep but when I reached over to awaken him, his body up-ended and I saw in a moment of horror that he had been cut in half just below the waist.'

The torture of the shipwrecked sailors lasted almost five days and nights. The men died of thirst, suicide or in the jaws of the sharks at the rate of about 100 every 24 hours. On noon of the fifth day a US airforce plane passed overhead. It continued on its course then suddenly swung about and came in low. They had been spotted! By the time rescue operations were complete at sundown, only 315 survivors were counted. In the years ahead many would be haunted by their nightmare experience in the Pacific Ocean.

Swimmers without life rafts would seem to have little hope when there are sharks in the vicinity. In a number of wartime instances servicemen with nothing to keep them afloat but their Mae West were shadowed but not savaged by the sharks. A US navy ensign who parachuted into the waters off Cape Engano in the Philippines during the carrier strike in October 1944 was one of those lucky ones.

'D. hit the water on his back. He received quite a jolt but was able to untangle himself and get out of his shoes and back pack. He found his life raft had shaken off when his parachute opened. The front half of his life jacket had a rip in it and the back half had to be inflated orally every half hour ... As he swam his socks gradually worked off, leaving his feet as a lure for sharks ... which promptly put in an appearance. He was shadowed by about four sharks 1.5 metres in length. They did not bother him as long as he continued to kick. As soon as he stopped to rest, one of them would make a pass at him. All of these were dry runs except one in which the shark grazed his legs and left tooth marks. He was picked up by a destroyer after 8 hours in the water.'

An Air Force C-124 crash landed in the Pacific between

Hawaii and Wake on 4 July 1958. With no life rafts the survivors got together a make-shift raft from the wreckage. There were nine passengers. Only three were picked up on the third day after the crash.

'About two hours after daybreak ... we had our first shark attack. We had seen them around, but now they were coming in close and making passes at our feet and hands. Whenever anything touched our feet, we would call out, "Who's that?" If no one answered, we would all try to climb on the small raft, which couldn't support one of us. I believe the splashing and kicking in this futile effort helped scare off the sharks.

'At one time, a shark came nuzzling in between Sergeant Vanderree and Sergeant Phillips and got himself tangled up in the mailbags. The shark panicked and so did the two sergeants. They finally managed to shove and kick him on through. The sharks averaged about 6 feet in length.

'About three hours after daybreak, the other courier began to talk incoherently and began floating away from the raft. He appeared to be actually swimming away rather than drifting. He possibly had thoughts of swimming to shore because at times he would ask, "Can't you feel the bottom?"

'It was on one of these "swim-outs" to get the courier that a shark grabbed me by the left shoulder and began shaking me like a dog with an old shoe. I guess he would have taken a chunk out of my shoulder if he hadn't wrapped his teeth around part of my shoulder bone. While he had me, we all started screaming, yelling, hitting the water and making all the noise we could. The shark let go, and I lost no time getting back to our raft.

'Every time the sharks would make an attack, which they did in packs, all three of us would try to climb on top of this little piece of wood. It was the proverbial "grabbing at straws" to get our feet away from the shark's teeth.

'The second courier was attacked and killed by the sharks. Our shark repellent was out, but we fast lost confidence in it.'

The sensational growth of air travel following World War II brought civilians into the realm of ocean survival. The possibility of crash was something no airline company wished to even mention. Regulations insisted passengers be given a brief 'emergency' talk by the hostesses at take-off. There were no handbooks issued about what to do if one crash landed in the sea — let alone the suggestion of shark attack.

None of the survivors of one of the worst postwar ocean

crashes knew the first thing about sharks. In November 1957 a Pan-American Stratocruiser crashed into the Pacific 1,600 kilometres east of Honolulu. There were 44 persons aboard. All survived the crash but 19 died from shark attacks. No one survived the Air France Super-Constellation crash off Dakar, Africa in 1960. All 63 lives were lost. It is believed many were taken by sharks. In February 1960 a small Portuguese airliner crashed into the sea between Timor and Darwin. All nine passengers were killed by shark attack.

CHAPTER EIGHT

▲▲▲▲▲▲▲▲▲▲▲▲▲▲

New South Wales

BEACHES AND RIVERS

The highest ideals of human courage have been achieved in the seas and the rivers of this continent. Anyone who claims Australian mateship is a myth should look at the record. The stories of the heroes who have plunged to the rescue of their mates and total strangers, without a thought for their own safety, have an important place in the folklore of this nation.

Time and again, at the risk of drowning, at the risk of being turned upon by the same killer already tearing into one victim, they have forsaken the safety of beach or boat, to come to the rescue. It might be said they acted more by instinct than by calculation. But surely a far stronger instinct is the one for self-preservation.

SYDNEY BEACHES

Coogee

One of the earliest rescue attempts which did receive justified recognition took place at Coogee Beach on 4 February 1922. It was not only one of the most spectacular and tragic shark rescues in Sydney surfing history, it was also the one which made the surfing public shark conscious.

There was a drizzle of rain falling and a good surf running on that humid and overcast Saturday afternoon. Coogee, at that time Sydney's second favourite beach to Bondi, was crowded. Someone caught sight of a solitary swimmer, some 35 metres out, flinging his arms wildly in the air. A thousand pairs of eyes turned as the water around the swimmer began to churn violently. Surf belt champion, Jack Chalmers, 27, of North Bondi was on Coogee Beach that afternoon. Like everyone else he knew exactly what was happening out there.

Coogee Bay, on 16 November, 1929, showing the shark proof fence on the end of the pier to the right

'If anyone ought to go out to that bloke,' he said to himself, 'it should be me.'

He raced across to a lifesaving reel. There was no belt so he knotted the rope around his waist. One of his mates manned the reel. Another let out the line. Eighteen-year-old Milton Coughlan was fearfully mauled by the time Jack Chalmers reached him. The water was crimson.

'At first I thought he must be standing on the reef because he was upright in the water and putting up a terrific fight. The shark kept rushing at him and circling away again.'

As Jack Chalmers came beside him, young Coughlan obediently flipped over on his back in accordance with rescue drill. It was a touching gesture. He had already lost an arm. Yet he even managed a few words to his rescuer, 'For God's sake mate, hang on to me tight.' He never spoke again.

Praying the shark would not return to attack, Jack Chalmers,

with the help of the rope, worked his way towards therocks at the southern end of the beach. The crowds of people on the sand watched tensely as a shark fin cruised back and forth, back and forth just beyond the two bobbing heads. Another man, Frank Beaurepaire, an Australian Olympic Champion, came thrashing through the water to join Chalmers and together they brought Milton Coughlan on to the rocks. The mob of spectators who came crowding round recoiled in horror at the sight of the shark victim.

Milton Coughlan died in Sydney Hospital just 25 minutes after the attack. A public fund was raised to reward Jack Chalmers and Frank Beaurepaire. With his £500 Beaurepaire began a national business. He was eventually knighted and became Lord Mayor of Melbourne.

One imagines scores of bathers wildly scrambling for the beach at the first sound of a shark alarm. In fact, overseas visitors are often open-mouthed at the nonchalant, almost reluctant manner in which an Australian crowd will wade to the shallows when the warning sounds.

The scene at Coogee Beach just before 11 am on the second day of March, 1922, however, was more in the nature of mass panic. The memory of the fatal attack on the young Newtown Railway station clerk, Milton Coughlan, on that same beach, was only 25 days old. Once again the seconds of disaster and distress were illuminated by the selfless courage of those who rushed to aid a fellow human.

Jack Brown fitted the ideal picture of the bronzed Aussie lifesaver. He was modest, dedicated; a fine physical specimen of a man. He was standing at the top of the steps when he heard someone in the distance call his name.

'Come quick Brownie!' He recognised Merv Gannon about 25 metres out. Gannon was further than anyone else. The water, still only knee-deep, was splashing all round him. Jack Brown could see the shark. He leapt down the steps and charged into the water screaming at the bathers, 'For God's sake mind! There's a shark!'

Women screamed and dragged their children to safety, others stumbled and splashed in the mass exodus to the beach as Mervyn Gannon tried to fight off the monstrous fish. It heaved at him and he punched, punched, punched it. Then it bit off his right hand.

Brown was beside him and acting as though the shark wasn't even there. He threw Gannon's left arm over his shoulder and half-dragged him towards the shore. A second man, Ern Carr, joined him. But the shark was not going to be cheated of its prey. It made its third and last lunge and tore a terrible gash down Mervyn Gannon's back.

First aid was administered on the beach and as they carried him to the ambulance he managed to utter a few simple words which seemed to sum up the fate of all shark victims: 'Well ... I have been unlucky.' Mervyn Gannon died of gas gangrene at St Vincent's Hospital.

Now Sydneysiders were aroused. Letters poured into the papers. Suggestions for observation towers were made. Submarines should be sent out with underwater cannons to blast all sharks to kingdom come. If not, then depth charges might do the trick. With World War I less than four years old, one returned soldier proposed the use of Mills bombs, used with good effect in the Somme Canal.

'The bait should be well placed and as soon as the sharks appear hurl the bombs at them. If the bombs struck, other sharks would come in after the blood and they would finish up eating each other.' The writer had 'no doubt there are plenty of willing diggers who understand the Mills, to give a hand to the job.'

The following Tuesday night at the dance held by the Coogee Boomerang Girls' Club in aid of a fund for Jack Chalmers who had gone to the rescue of Milton Coughlan, half the proceeds went to another hero who shared his first name — Jack Brown.

Bronte

It was dusk when Nita Derritt, 30, of Duntroon Street, Hurlstone Park, waded out from Bronte Beach on Wednesday 13 February 1924. In company with her relatives, the Browns of Waverley, she was laughingly splashing herself in waist-deep water. What happened next occurred in the space of one minute. Her left leg was severed below the knee. She thrust out her hand to fend off whatever it was under there. In its second lunge the shark almost removed her right foot.

She was helped ashore by her two young nephews and by a policeman, Constable Rushbrooke, who was on the beach at the time.

Miss Derritt showed incredible fortitude. Although shockingly injured and in great pain, her only concern was for her 70-year-old mother. 'Don't let mother see,' she gasped. 'Don't let mother worry.' She even managed to flick a towel over herself just in time and left the beach for St Vincent's Hospital, with her mother still believing she was suffering from severe cramps.

Nita Derritt was employed as a saleswoman at Saunders', the Sydney jewellers, for more than ten years. Eighteen years later another young woman employed by a Sydney jeweller would be mauled by a shark.

On the following afternoon, the *Daily Sun* reported that her right leg had also been removed. Her condition was serious. It also stated, 'Mr Saunders drove straight to the hospital from his home in Darling Point this morning and expressed the desire that no expense should be spared in caring for the unfortunate girl.'

Nita Derritt never saw the shark that made her a cripple for life. 'I did not know until afterwards that I had been in the horrible grip of the shark.' Because of this she was spared the tremendous shock to the system which victims experience in a shark attack, often a major contributing factor to their death. Perhaps this is why Nita Derritt survived while others with less serious injuries died.

But there are some who will never learn. The very next day, members of the Brighton-le-Sands Surf Lifesaving Club, about 14 kilometres from Bronte, applied to the Randwick Council for spotlights for the beach so that bathers could enjoy surfing at night. The aldermen were adamant. They refused.

Shark lookouts were used in pre-W.W.I beaches in Sydney's Eastern Suburbs. When the bell rang the board saying 'Shark' was displayed.

Mrs Dagworthy did not want her son to go to the beach. It was late in the afternoon and tea would soon be ready. But 16-year-old Jack Dagworthy, proud to have become a recent member of the Surf Lifesaving Association, wanted one quick dip. After all, Kurrawa Avenue, where he lived with his sister and recently widowed mother, was a hop-skip-and-jump from the surf. He would be home by six.

There was a nip in the air at 5.30 at the tail-end of a fine warm day, 27 March 1925, when Jack Dagworthy entered the surf at Coogee to become the fourth and last victim of a series of attacks along the Coogee shoreline which began in February 1922.

'I had no chance. He was on me before I could move. I heard the shark bell. I didn't see the shark and I felt ...' He was waist-deep in boiling foam when the shark clamped his leg and tried to drag him into deep water. 'As soon as the shark grabbed me I offered a prayer to God. It let go and swam away.'

He staggered ashore with his left leg badly mutilated. Blood transfusions were administered at St Vincent's Hospital. His condition was serious. Doctors decided on an immediate mid-thigh amputation. When he came out of the anaesthetic he caught sight of his mate, Tom Irwin beside the bed. 'Tom,' he said. 'Whatever you do, never get bitten by a shark.'

Jack Dagworthy recovered. Perhaps his prayer was heard.

Bondi

Bondi Beach, that magnificent sweep of sand that curves around the blue Pacific a mere 8 kilometres from the concrete heart of the city, has attracted hordes of sun worshippers for 60 years.

There have only been three shark attacks at Bondi. The first was in 1928. The other two, both fatal, occurred within 14 days of each other, early in 1929.

April is late in the season for both swimmers and sharks. But on the 14th of that month, in 1928 Max Steele, 19, was one of a group of lads surfing 100 metres from the sands of North Bondi Beach. The young surfers picked their wave and it sent them shooting into the shallows. All except Max Steele. When the next 'beauty' came riding down on him, Max, the last man out, caught it on the curl. At that moment he felt a sharp, stinging pain in his left leg. He never saw the cause. When he crawled on to the sand his leg was a bloody mess. His mates applied a tourniquet and raced him through the Saturday traffic to St Vincent's Hospital.

He lost his leg but remained alive never to be 100% certain that it was in fact a shark that had made him an invalid. Colin Stewart, 14, who lived across from the beach at 22 Campbell Parade was not so lucky.

In January 1929 Sydney played host to the celebrated

Olympic swimmer Arne Borg. On the second Sunday of that month, Colin Stewart, a tall, strapping lad, was among the spectators who packed the Domain Baths to see the Swedish champion give a series of demonstrations.

Members of the Bondi Surf Lifesaving Club in 1929

Colin arrived home for tea at 5 pm and no doubt inspired by what he had seen that afternoon, went across to the beach after he had eaten. The surf lifesaving patrol had knocked off ten minutes before, at six o'clock, but the two flags were flapping in the late afternoon breeze. In spite of the dull weather and a choppy sea, 60 to 70 bathers were shouting and splashing between the flags when Colin joined them. He ducked under to wet himself and kept wading out.

He was 18 metres from the shore. There were several heads bobbing about still further out. The voice of beach inspector Ken McKenzie sounded over the din as he bellowed at a group of bathers to keep within the flags.

Robert Kavanaugh who was the first to grasp the boy and helped to bring him in. On reaching the beach he disappeared.

The shark made two savage lunges at Colin Stewart. It tore an enormous gash 125 millimetres deep from thigh to hip. Two men, Kavanaugh and Kelly, plunged to the rescue. The surf patrol raced out across the sand. Colin Stewart smiled gamely at Jack Kelly as he lowered him gently to the sand. Then he passed out.

Club members packed towels around the gaping wound and the Eastern Suburbs ambulance clanged through to St

Vincents. A young member of the North Bondi Surf Club called at the Hospital later in the evening to offer his blood. Mrs Stewart thanked the boy through her tears. Her son was already dead.

Just under four weeks later at 4 pm on a showery Friday, 8 February 1929, a 39-year-old man was attacked at Bondi. He was among a group of people swimming in 3.6 metres of water when he was singled out by the shark. He was John Gibson, a son of the owner of the Melbourne department store, Foy and Gibson. In spite of prompt medical attention on the beach, he died of shock and haemorrhage before reaching the hospital.

The two fatal attacks at Australia's most popular surfing beach came as a shock to the community. A. E. Whitelaw, the British soap magnate who was in Australia at the time, offered £1000 to the President of the Surf Lifesaving Association, C. D.

Paterson, to erect a test shark fence. The North Bondi Lifesaving Club who had been trying to get a boat for some time, were soon presented with one. Yet in spite of the fact that anything like real

Board-riders freeze 274m off North Bondi. The shark circled the boys for 40 fearful minutes before a launch chased it out to sea.

protection from sharks was still eight years away, the surf at Bondi remained as crowded as ever.

Maroubra

At times the mysterious workings of fate can be glimpsed behind the incidence of a shark fatality. So many times the victim has gone in for 'one last swim' or 'just one more shoot', never knowing the significance of those words.

Four young men were together in the surf at Maroubra, one

of Sydney's largest southern beaches, on 18 February 1929. It was high water time that afternoon, and close to what has since been recognised as peak shark attack time.

Ritchie Dunn and Allan Butcher were together riding the waves. They came tumbling into the shallows together. Dunn had his share of fun and rose to his feet saying he was going back on the sand. Allan Butcher said he would catch 'just one more.' He was now beyond the front line of bathers, about 60 metres from the shore. Suddenly Butcher screamed out, 'Billy, Billy, oh Billy!'

William Harrison who was closest to him, swam frantically against the waves towards the stricken man, closely followed by Harry Clay, the fourth in the group. The shark which Butcher had been trying to push off with his hands and feet fled and that 'last shoot' carried Butcher's body some 35 metres towards the beach.

He died of wound infection at Prince Henry Hospital. His friend, Richard Dunn, who had gone ashore and whom fate had spared, remained in the hospital with Allan Butcher to the end.

One heroic effort to snatch a victim from the very jaws of a shark was made by a 17-year-old unemployed youth at Maroubra in 1935.

Saturday 9 March was a day loaded with dangerous possibilities. It was a sultry day, cloudy, with a drop of rain in the air. Earlier in the morning a school of sharks had been spotted cruising north of the Harbour, between Newport and Mona Vale. Ernest MacDonald, 27, decided to go down for a dip just before lunch. He left his wife and young toddler at home in Bay Road, Maroubra and strolled to the beach, his towel over his shoulder, wearing a pair of red trunks. There were a fair number of people on the beach. A dozen or so were bathing between the flags at the southern end. There was a good surf running and a small inshore channel. The water was warm.

Only minutes after Ernest MacDonald entered the water, his mangled body was brought ashore. He was 25 metres out in chest-high water when the startled bathers heard his first scream. Scores of people ran to the water's edge as they saw the man beating off a shark with his hands, disappearing and reappearing as the shark dragged him under. Without hesitating, a Randwick lad, Bill Wright, rushed through the water to MacDonald's aid.

'I grabbed him by the arm,' Bill Wright said later. 'Then the shark came at me. It missed me and got MacDonald by the other arm. Its teeth brushed by my fingers. I felt a terrible pull as the shark wrenched at MacDonald's flesh and exposed the bone. It swept by us and returned almost immediately with incredible speed. I yelled to the others to take him in.' Members of the Maroubra Surf Club came to his assistance.

'I splashed and shouted as hard as I could, but still it kept following us, even when we were only in a few feet of water it pursued us. By the time we got MacDonald ashore he must have been drained of blood for a long crimson trail stretched far out to sea.'

When the Eastern Suburbs ambulance arrived, Ernest MacDonald was dead. There were bloodstains everywhere, on the sand, on the promenade. The shark was seen cruising up and down in the shallows for a full 15 minutes after the attack. It was about 4 metres long. The following day Ernest MacDonald's brothers called at the Maroubra Surf Club and solemnly thanked the members for their endeavours.

North Brighton

To three youngsters, Max Farrin, 13, Ken Moore, 12, and Harry Flower, 11, on the morning of 23 January 1940 the European war was remote and somehow heroic. They were enjoying the break from school as they skipped across General Holmes Drive, their towels under their arms on their way to North Brighton Beach for a swim in the waters of Botany Bay.

At about 10.40 am a man, sunbaking on the sand, was approached by a wild-eyed boy who said his friend was covered in blood in the water. The man jumped to his feet and could see the boy, Max Farrin, floating 45 metres out. He hurried into the water, then hesitated. He had seen the black fin cutting the water close to the prostrate boy. He turned and ran back, stumbling up the sandhill and on to the road where he hailed a passing truck to obtain some rope to get the boy out.

Syd Owen, whose house near the corner of Bestic Street faced the spot where the boy was drifting, had been peacefully watering his garden. Attracted by the disturbance, he promptly dropped the hose and raced over to the beach. Someone yelled that the shark was still there. Syd Owen did not falter. 'I waded out and splashed about until the sea was beyond my depth. Then I swam to the boy who was floating in about twelve feet of water. When I caught hold of him I could see he had been shockingly injured.'

Spectators on the beach held their breath. The shark was

no more than 3 metres from the man and boy. Syd Owen saw it too. 'Must have been at least 10 feet. Luckily it didn't attack again.'

Max Farrin was wrapped in blankets and rushed to St George Hospital where he died of massive leg injuries soon after. Australians everywhere gave a mental salute to Sydney Owen, a former Digger with the 54th Battalion, wounded in the arm in France 1918 in the war to end all wars. Syd Owen was wrong about one thing. The shark did strike again — only 12 days later and 400 metres away. Here too, was to be the scene of a brave rescue.

John Eke had arrived in Australia from England in the late twenties. He was a solitary man in his mid-fifties who lived at a Salvation Army Hostel. He worked as an accountant. It was a hot day with an early afternoon temperature of 30°C. To protect his pale skin from the glare, John Eke kept his white shirt on over his costume. At 2 pm he was in waist-deep water about 12 metres from the beach. No one saw anything unusual happen. Perhaps it was his innate independence that prevented John Eke from crying out. No one quite knows what happened. He began to swim for the shore, his arms torn from two assaults made by the shark, trailing behind a ribbon of blood.

Bill Kennington of Arncliffe and his 14-year-old son Wallace were standing on the beach when they heard someone shout, 'Shark!' Without hesitating, father and son ran into the water where 10 metres from the shore in 1.5 metres of water John Eke was still making a feeble effort to reach the beach. The pair of them half lifted, half dragged him to the shore.

Wallace Kennington's boy scout training proved useful as he helped his Dad try to save John Eke's life. They made tourniquets and tried to staunch the flow of blood. The stricken man was already unconscious.

'We didn't stop to think about the danger,' Bill Kennington said later. 'We did our best to save the life of the man.' In spite of their gallant efforts they were not successful. John Eke died from shock and loss of blood at St George Hospital some hours later.

Northern Beaches

Perhaps because the beaches north of Sydney Harbour were the last to be settled, there is no record of shark attacks, right up to January 1934. The spell was broken after that. In the next two years there were five attacks north of Manly. Only one, the first victim, escaped with his life.

Sunday 7 January 1934 was not exactly the sort of day that would attract many people to the beach. It had been raining, sometimes quite heavily, on and off. The water off Queenscliff beach wasn't particularly inviting. The rain had sent out drainage from Manly Lagoon and the water was muddy. But it did not deter Colin Grant, 22, and a couple of his friends. They were way out, 35 metres off shore, on a sandbar, the water up to their shoulders.

Suddenly Colin yelled, 'Shark!' and disappeared. His mates laughed. They thought it was a joke, like crying 'wolf.' He reappeared and screamed again. They helped him across the deep channel towards the shore and already the alert lifesavers were rushing to meet them with belt and line. While still in the water they gave him a tourniquet. This prompt action probably saved his life and fortunately Manly District Hospital was only a few miles away. Colin Grant lost his right leg and recovered in spite of gas gangrene infection.

It is said that three laughing Fijians plunged into the deserted surf soon after the attack. They waved aside warnings about the shark. Fortunately this was one time when the myth that a black skin is less likely to attract a shark than white was borne out.

Nine weeks later and several kilometres north of Queenscliff at Dee Why, at 3 o'clock on a Monday afternoon, Athol Riley, 17, was one of a group of bathers well within the breakers and in water hip-high. Suddenly a shark fin appeared among them. With little more than a gasp Athol Riley went under. He reappeared a split second later before the horrified view of the bathers. His leg was in the shark's mouth. There were screams and a rush for the shore. The boy's leg and buttock were torn off and he died moments after three men got him to shore. A 4 metre shark cruised up and down for two whole hours and repeated attempts to catch it failed.

Twelve weeks after Colin Grant was rushed into Manly District Hospital, 15-year-old Leon Hermes was carried through the door of the outpatients' department on 1 April 1934. The staff did not have a chance to save him for he died minutes after admission.

Leon had been in the front line of breakers in waist-deep water, 7 metres from the sands at North Steyne. Once more rainwater from Manly Lagoon had discoloured the surf. The shark made three strikes. Rescuers used a lifeline as a tourniquet and packed towels to the wounds to his right leg and thigh. Once more the shark stayed in the area after the attack.

In the thirties one could take a rollicking tram ride, with sweeping views of surf and distant headlands and a sea-breeze blowing, from Manly over to Narrabeen. Beyond the terminus, straggling rows of holiday cottages stretched as far as North Narrabeen. A dozen or so people were in the water which was particularly warm at 24°C when Herbert McFarlane, 22, splashed in for a last minute dip at 5.30 pm. The water where he stood was shallow, 90 metres from the other bathers.

The attack was ferocious. The shark came at Herbert McFarlane repeatedly and even when his rescuers, John Barrett and Carl Read, carried him out, the shark, 3.5 to 4 metres long followed the blood trail to the shore. Herbert McFarlane died from a massive wound to the thigh before he reached hospital.

There have been no further fatal attacks on a northern beach beyond North Narrabeen since that second day in March, 1935.

A beach inspector, Dudley Beer, at South Steyne at the southern end of Manly Beach, figured large in an impossible rescue attempt. More than 100 people were splashing and laughing in the surf on a hot and sticky Tuesday afternoon, 4 February 1936. David Paton, 14, and several other swimmers were just beyond the main body and 25 metres from the rocks. An onlooker saw a huge dark shape rise out of the water and collapse on top of the boy. There was no sound, just a big splash. Beach Inspector Dudley Beer was first to the spot but there was no sign either of the boy or the shark. There was only a large bloodstain in the water.

Young David Paton was the last person to die from a shark attack on a Sydney surfing beach. Twenty months later in October 1937, meshing was introduced off the metropolitan beaches.

The record was almost broken on 6 December 1975. A young man was observed, frantically paddling his surf ski towards North Steyne Beach with a black dorsal fin a metre behind him. The man was in an understandable panic, screaming and yelling at the top of his voice. Fortunately the surf power boat was in the vicinity. It bounced across and cut in between the surf ski and the 3 metre shark.

'It was touch and go,' said a spokesman for the Shelley Beach Rescue Unit. 'He was that scared, even when he got to the beach he just kept on going.'

Meshed Beaches

North and South of Sydney Harbour

North

1. Stockton
2. Nobby
3. Newcastle
4. Merewether
5. Redhead
6. Bar
7. Swansea
8. Caves
9. Catherine Hill Bay
10. Palm
11. Whale
12. Avalon
13. Bilgola
14. Newport
15. Mona Vale
16. Warriewood
17. Narrabeen
18. Dee Why
19. Curl Curl
20. Harbord
21. Queenscliff
22. Manly

South

23. Bondi
24. Tamarama-Bronte
25. Coogee
26. Maroubra
27. Cronulla
28. Austinmer
29. Thirroul
30. North Wollongong
31. Wollongong

In 1988 there was a storm of public protest when the NSW Government discontinued meshing of beaches in August and May, reducing to eight months the protection which had previously been in place from 1 August to 31 May on NSW beaches north from Sydney to Newcastle and south to Wollongong. The cost cutting measure was denounced as showing that the government had no concern for the dangers to swimmers and surfers in the winter months. There had been a disturbing rise in the number of sharks netted off Sydney beaches in the preceding years. The protests went unheeded and in July 1990 a Sydney surfer had 'the fright of his life' when a large shark cruised past his board at Bondi Beach.

However an opposing view was now more frequently heard that the advantages of beach meshing were outweighed by the large numbers of harmless marine creatures which became entangled in the nets and died. Many people believe that spotter planes would provide an equally effective and more humane form of protection.

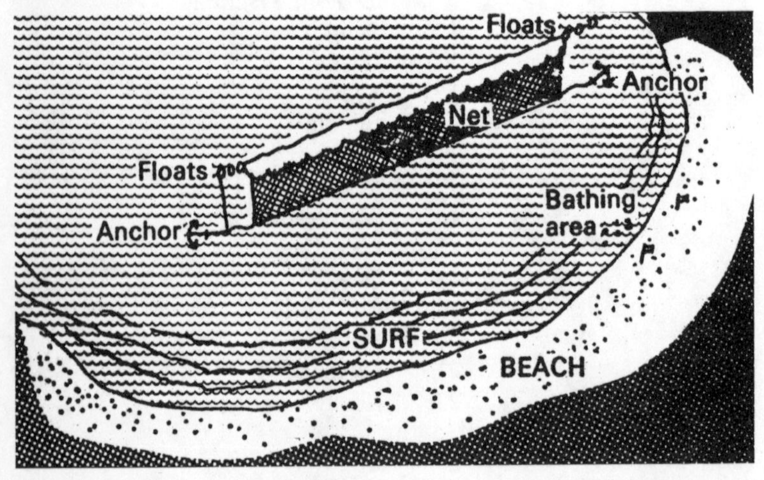

Caught in the mesh. Diagrammatic view of meshing operation. Once a shark runs into the net it is doomed. Instead of retreating it presses on in a bullheaded attempt to force its way through.

SYDNEY HARBOUR

An item in the *Sydney Gazette* dated 20 July 1806 headed 'A Caution to Parents,' advised them to 'keep their children away from the hospital wharf where a large shark had been seen cruising.' Considering the amount of activity that has taken place on Sydney Harbour going on for 200 years it's a wonder the records do not show a long list of shark attacks. As the paragraph in the Gazette indicates, Sydneysiders have been conscious of their Harbour sharks from the earliest days. Names like Shark Bay, Shark Island and Shark Point bear witness to this.

In the history of the Harbour the sharks have been given some fine opportunities which, fortunately, they did not take up. In the wrecks of the *Dunbar* in August 1857, and the two ferries *Greycliffe*, November 1927 and *Rodney* in February 1938, none of the survivors who floundered in the waters of the Harbour were approached by sharks. There appear to have been several fatal attacks through the 18th century but records are scanty. There were two fatal attacks in Sirius Cove, one in 1913 the other in January 1919, when 13-year-old Richard Simpson was attacked while wading close to the shore.

New arrivals from Britain and elsewhere are often wide-eyed at Australians' seemingly blasé attitude towards sharks. One new arrival had no fear of sharks and it eventually cost him his life. Walter German was an Englishman, employed as a motor mechanic, in Sydney in 1916. He and his wife built themselves a cottage in Buckhurst Avenue, Point Piper. They were both keen swimmers and regularly, each morning, went for a dip in the Harbour off Seven Shillings Beach, before going to work.

On Friday 8 December 1916, Walter German had the day off and he and his wife took their swim later in the morning. About 10 metres from the shore, Mrs German was nipped on the leg by a shark. She screamed and her husband plunged out to her. Together they made for the shore. Only a metre from the shore the shark attacked Walter German. His wife tried to fight it off. But it punctured his chest and took his right arm and he died with his wife still clinging to him.

One bizarre incident was reported on 9 April 1918. A man named Ansell, aged 28, a ship's fireman, raced along the deck of the Parramatta River ferry steamer screaming. 'They're after me!' He dived overboard and began swimming for the shore. He had gone no more than 10 metres when dumbfounded passengers saw a shark fin cleave the water in his direction. The man screamed, his sanity perhaps returning. The shark appeared to back off in fright and swam in a circle lashing the water into a foam with its tail. The watching passengers shuddered. It kept returning to the attack, but the man fought frantically, hitting the shark on the nose. Then the shark came up under him, lifting him bodily out of the water. A launch came to the man's rescue. Apart from severe shock he sustained no injuries.

MIDDLE HARBOUR

At the beginning of 1942 came the first of six devastating attacks strung across the years that were to make Middle Harbour the most notorious area for shark attacks in the whole of Australia.

Zita Steadman, 28, single, attractive, private secretary to W. J. Proud of Prouds, the Sydney jewellers, had everything in front of her when she left her Ashfield home on that hot, hazy Sunday morning, 4 January. In company with her neighbours, Mr and Mrs Bower and Mrs Bower's mother, Mrs Reeve, they arrived at a boat hire shed in Middle Harbour. There were several servicemen and their girlfriends around for it was wartime 1942. The picnic party chugged up Middle Harbour to a spot near Bantry Bay known to fishermen as Egg Rock, 6 kilometres from the old Roseville Bridge. They tied up the hired motor launch and spread their picnic hampers on a flat rock ledge.

The water was crystal clear and deliciously refreshing as Mr and Mrs Bower and Zita Steadman waded in. Close to the shore the water in this spot is shallow. The sandy bottom slopes down gradually for 7 metres then drops steeply to 9 metres. Zita was 3 metres beyond the husband and wife, waist deep. Mr Bower called, 'Don't go out too far!' She began to wade back towards them. On the rock ledge Mrs Reeve scrambled to her feet and opened her mouth to cry out a warning as she caught sight of the shark. Simultaneously, Miss Steadman screamed and threw up her arms.

Mr Bower grabbed an oar from the boat and slammed it against the shark which was repeatedly attacking the girl. It made no difference to the 4 metre long beast. It was dragging its swooning victim into deeper water. Bower threw himself into the launch, started the engine and tried to ram the shark as it leapt at its prey, lashing its tail with ferocity. This failed too. He cut the motor and jumped into the water. He seized Zita Steadman, now mercifully dead — by her long black hair. The ghastly tug-of-war between the 4 metre shark and the 95 kilogram man was brief. The body was torn asunder. The shark slipped back with its prize into the murky depths.

While these terrible events were being enacted, round the point, out of sight, two RAAF men and their girlfriends were laughingly dangling their arms and heads over the side of their boat, to keep cool.

Boxing Day 1942 looked like being a bonzer day, just as

Denise Burch and her elder sister Pam, had been hoping all Christmas week. There were eight of them, four boys and four girls. The eldest was eighteen. At 15, Denise Burch was the youngest. The sun was already hot at nine in the morning when the party gathered at the Spit boatshed, joking about who would go in which of the two boats. The two boys in each boat took the oars and there was some talk of a race, which didn't come off. Instead they moved leisurely through the crystal waters of Middle Harbour — they were still crystal in 1942 — towards a point near Bantry Bay.

At ten to eleven two of the boys were swimming in deep water 20 metres out from the shore. Denise was happy to cool off, paddling in half a metre of water. Then she screamed. She was seized about the legs and toppled backwards. The shark tried to tug her into deeper water. It was almost a carbon copy of the attack on Zita Steadman at the beginning of the year.

Those on shore grabbed for the nearest branch or rock they could lay their hands on, and rushed into the water pelting the shark, clearly visible through the cloud of blood. Denise reappeared, floating face down on the water. The shark had fled before the barrage.

Denise Burch had been evacuated from Hong Kong with her mother and sister (her father remained as a prisoner-of-war) two years earlier. Danger is all around, ready to reach out and touch us when we least expect it. Denise was already dead when they gently laid her body at the water's edge.

On 14 January 1947, two youths, Bob Scott, 17, and John Blackwall, 16, were on a row boat on Sydney Harbour and had turned into Berry's Bay, just north west of the Bridge, when they had the shock of their lives. Their boat was suddenly lifted nearly half a metre out of the water. They were in a hair's-breadth of capsizing and landing in the drink. As they splashed down again, a 4.5 metre shark emerged from beneath their boat, made two circles around it and then disappeared. 'We had the shakes properly. We rowed as fast as we could to the shore.'

The word flew from mouth to mouth faster than the fastest bushfire. 'Shark! Someone's been taken by a shark.' All along Edwards Beach, a boomerang-curve of sand separated from Balmoral Beach by a rock promontory, people scrambled to their feet and shaded their eyes.

On that Tuesday afternoon, 18 January 1955, Mrs Willis of Balmoral was sitting on the warm sands watching her young daughter playing. Although she knew her 13-year-old son was in the water looking for lobsters, so remote are the chances of shark attack, it never occurred to her it might be her John who had been taken. Her intention, as she hurried, with a number of others to the spot known as Wy-ar-gine Point, was to give help.

She edged her way through the crowd and recoiled in horror when she saw her son lying unconscious. His legs and his arms were covered with towels. A trail of blood ran from the rocks over into the sea. John Willis had been attacked in 2.4 metres of water. In spite of deep incised wounds to his right calf and skin abrasions to his right hand, John Willis managed to make the shore. He died from haemorrhage.

John was a pupil at Mosman Intermediate High School. His death, editorialised in the *Sydney Morning Herald*, was 'a grim reminder of the danger which always lies beneath the smiling surface of Sydney's bays and inlets.'

Saturday 5 February 1955, less than two weeks following the attack on John Willis, was a warm 26oC. Two men, covered in mud and perspiration, were cleaning out a shark-proof rock pool in Sugarloaf Bay. They finished the job about 2 pm. One of the men, Bruno Rautenberg, a 25-year-old German migrant, kicked off his boots. 'I think I'll have a swim and wash off the muck.' His companion, Kenneth Wood, in whose Harbourside home Rautenberg boarded, shook his head. John Willis had been fatally attacked 5 kilometres away. 'Wouldn't it be best to wait until the pool filled?' Rautenberg laughed and dived off the wall with a loud splash into 4.5 metres of water.

Kenneth Wood shrugged and began walking across the lawn back to the house. He heard Bruno Rautenberg scream. Believing it was a joke, Wood nevertheless walked over to the wall. There was no one in sight. The ominous silence was broken by the waves slapping against the rocks. Suddenly a man's agonised face broke the water and he gave a piercing scream. For a moment he thought Rautenberg was drowning. Then the sun flashed on the white underbelly of a shark. Wood was frantic. He could not swim. 'I ran to the bank near him. There was blood in the water all around. The shark was right beside him. I saw it grab him by the leg, drag him under and hang on.'

Wood spotted a 3.5 metre piece of water piping. He

grabbed it and thrust it out at the shark. It appeared to distract the fish. It released its hold. Rautenberg began swimming lamely towards the rocks. He covered 9 metres when the shark was upon him again. Wood, clutching a garden rake, waded out several feet and began hitting the shark with the steel prongs. With his other hand he grasped Rautenberg, fending off the shark with the rake. 'He was dead when I got him ashore.' For some hours after the police and ambulance had left, the 3.6 metre shark 'with a sharp pointed nose and black along the back' continued to cruise the area.

Why do swimmers take risks with sharks? For the most part it is their confirmed belief in their personal immunity from attack. It could never happen to them; someone else perhaps, but never them. For others the potential danger is, in itself, part of the attraction. The effect on them is not unlike that experienced by a gambler. Some do it out of sheer bravado or to show off their fearlessness to others. We will never know which of these motives prompted Bruno Rautenberg to swim in Middle Harbour that afternoon. It certainly was not the first time he had done so. In the two months he had been living with Mr and Mrs Wood he had, in spite of their pleas, swum 130 metres across the bay. As a Christmas feat, several weeks earlier, he had swum about 800 metres across Middle Harbour. Rautenberg's wife and daughter were still in Germany. She had been hesitating about joining him in Australia, in spite of his glowing accounts of the country. She would never see Australia now.

These attacks were followed by the biggest shark hunt ever witnessed in Sydney. Almost everyone who had ever hooked a shark was out in every imaginable craft. Philanthropist Sir Edward Hallstrom offered a purse of £1,000, half to the fisherman who captured the killer shark and half to Chatswood Community Hospital. Nicholas Gorshenin, the shark meshing contractor, set nets across Sugarloaf Bay.

On the day following the attack, Sunday, two teenage lads, Bill Harvey and Bruce Graham were among those who went out 'to get' the killer. With some of the earliest scuba gear, the boys waded into the waters off Balmoral Beach. Neither of the pair quite knew what they would do if they come face to face with the killer shark, but they were as stirred up as the rest of Sydney's citizens.

About 40 metres out a dark shadow began to circle them.

They froze. It was a 1.8 metre shark. Remembering the maxim to keep one's eye fixed on an inquisitive shark at all times, the boys stood motionless, back to back, while the shark kept on swimming around them. Then slowly, maintaining that position, they began making sidesteps in perfect unison back towards the beach. The shark, no doubt puzzled by this two-headed monster, decided it might best be left alone and promptly disappeared. The lads continued their sidestepping back to the shallows and left the shark hunt to those better equipped.

In all, nine sharks were caught that week. Radio personality Bob Dyer, at the peak of his popularity at that time, caught a 350 kilogram tiger. His wife, Dolly, caught a 220 kilogram bronze whaler off Ben Buckler. There was no evidence to prove any of the sharks was the one that attacked either John Willis or Bruno Rautenberg.

Theo Brown was a diver with the Northern Territory Police Force in 1960. That Christmas he was on leave at his parents' home on Sydney's north shore. He quickly made friends with his Forestville neighbour's 13-year-old son, Ken Murray. Ken was an intelligent, adventurous youngster 'a real Huckleberry Finn who was always keen on the water.'

During his boyhood in the late 1940s, Theo Brown used to swim at a spot near the old Roseville Bridge in the upper reach of Middle Harbour. The water then was sparkling clear; now it was murky with poor visibility resulting from refuse lying on the bed.

Brown and Ken were skindiving and a youth who accompanied them was on the shore. The two were in 1.2 metres of water when, without the slightest warning, Ken Murray was attacked. Theo Brown, in spite of his dismay, acted promptly and efficiently. Ken's leg was badly mauled and he was bleeding profusely. Brown got him to the shore and applied a tourniquet. There was a dramatic rush to the Royal North Shore Hospital. Ken's condition was critical. For days his life hung in the balance. Theo Brown, tormented with self-recrimination, although completely exonerated and indeed praised, by the parents, remained at the hospital the whole time.

Ken's leg was amputated and he appeared to rally. Towards the end of the first week he took a sudden turn for the worse. On 25 January 1960, nine days after the attack, Ken Murray died. His classmates from Chatswood Boys' High School formed a guard of honour. Deeply moved by the tragedy, Theo Brown vowed he would give some meaning to Ken Murray's untimely death. He has since devoted all his time in search of a shark repellent that would prevent such things happening again. He has carried out research for the Americans and for the French in Tahiti.

On that same Sunday, while young Ken Murray's life was ebbing away at the Royal North Shore Hospital, a Sydney Morning Herald reporter visited some of the popular beaches in Middle Harbour. At Balmoral bathers outnumbered those in the shark-proof enclosure. It was the same story at Edwards Beach with at least 100 people, mostly families, outside of the safety enclosure. Further round at Wy-ar-gine Point, several fishermen stood in waist-deep water.

MIDDLE HARBOUR ATTACKS

1. Zita Steadman, 1942
2. Denise Burch, 1942
3. John Willis, 1955
4. Bruno Rautenberg, 1955
5. Ken Murray, 1960
6. Marcia Hathaway, 1963

NSW — OUTSIDE SYDNEY

South Coast Beaches

In the summer of 1966 Ray Short, 13, was staying with his parents at a holiday cottage at Coledale, a straggling ocean beach resort, 50 kilometres south of Sydney. The afternoon of 27 February was hot and still. Ray decided to cool off in the surf. He plucked his bathers from the line and plunged boldly into the marvellously refreshing waves. The event which followed describes that moment, in all the exhilaration of surf and sun and the joy of being alive, when one is abruptly faced with the awful prospect of death. 'From the moment I saw that black shadow I knew I was going to die. I knew I would never see my Mum and Dad again or my mates.'

The surf was ideal that afternoon and Ray was having a great time. He was 25 metres from the shore and treading water when he decided he had had enough. Another majestic wave came rolling towards him and he hesitated. 'Just one more.' He felt a nudge on his left leg. He glanced over his shoulder and saw something dark in the water.

Coledale on February 27, 1966, showing lifesavers examining the shark which attacked Raymond Short.

'I thought: "Oh my God. Please no. I'm going to die. It's a shark." It shook my leg then let it go and swam around me in a circle. I wondered why there was no pain. I could see the blood. The water was turning pink. It hit me again. Another bump just like the first, only this time on the right leg. My whole body seemed numb from the waist down. I thought: "The shore. If only I could reach the shore. I could make it." Suddenly I was lifted up out of the water. The shark had my whole leg in its mouth, its jaws just over my knee.'

Ray tried to dig its eye. 'Up till now I hadn't really seen the shark. But I found myself staring it right in the eyes, the hideous, staring eyes. His teeth seemed to be grinning at me.'

In his desperate bid for life, Ray bit the shark on the nose; a good, hard, long bite. 'It's nose was like hard, old, salty canvas.' It had no effect on the shark. 'I could taste my own blood in the water. The whole sea was red. Then my right foot felt something hard. Now I was standing on one leg. I had been washed on to a sandbank.'

The shark went limp but it still hung on. There were voices. 'It's okay mate! She'll be okay. The shark's gone.' But the shark had not gone. 'I kept saying: "It's still there!" They kept telling me it had gone. They'd have me out in a flash. Then someone ran their hand down my leg and screamed. "By God he's right! The bloody shark is still there!" ' The blood-red water had hidden it from view.

They carried the 2.5 metre shark and the boy up to the beach. It still would not let go his leg. They bashed the shark with a surfboard. They beat it with sticks. One man rushed off to fetch his rifle. Then the shark unlocked its jaws. Ray Short spent many months in bed. He underwent 14 operations. Happily he lived.

Wollongong boasts the third largest population centre in NSW. The district, 80 kilometres south of Sydney, is fringed with miles of cream-coloured sands. Thousands bathe there all through summer. There is no shortage of sharks. Hundreds have been spotted and dozens have been caught. Yet in Wollongong's entire history there is no record of a fatal shark attack.

Central Coast

The Inman family of Concord, Sydney, decided to spend Christmas 1934 on the edge of Brisbane Water. The holiday cottage nestled into the foot of a steep wooded hill on Horsefield Bay, 1.6 kilometres from Woy Woy, about 50 kilometres north of Sydney. They arrived, full of anticipation, on Friday 21 December.

A number of young people from neighbouring cottages had been swimming with the two Inman children, Ray, 14, and Joyce, 12, in 3 metres of water, most of Sunday morning. Everyone was at lunch. Brother and sister finished theirs quickly and at 1 pm were ready for more fun in the bay. Their older sister, Kathleen, 26, said she would slip into her costume and join them.

Mrs Inman relaxed in a wicker chair, in the shade of the

Searchers trudge along Wamberal Beach, NSW *in* March 1955 *hoping to a find trace of missing* Noel Langford. His *torn swimsuit was washed up on the sand four days later.*

porch, watching her two children engage in a diving competition off the end of the tiny jetty. Who could make the least splash! Joyce dived and Ray, shouting he could do better, followed her. Almost the instant he broke the water, Joyce screamed as a large black fin came towards her. She kicked out and something sharp grazed her leg.

Mrs Inman sprang to her feet in horror and screamed, 'Shark!' Kathleen appeared at the door and saw her young brother vanish in a swirl of foam. She raced over to a boat and rowed across as Ray momentarily surfaced. The shark struck again and the boy went under. They never saw him again.

The boat was right over the spot, its timber stained red. Desperately Kathleen Inman thrust below the surface of the water with the oar. Immediately she came ashore, she rushed to the Wasson family next door and warned them against allowing their children to swim in the bay. Police dragged the area until 8 o'clock. Later in the evening the distraught family returned to their home in Concord.

On Sunday morning 13 March 1955, a man strolling along the beach near Wamberal, 10 kilometres from Gosford, NSW prodded a piece of rag washed up from the sea, with his toe. He saw it was a torn pair of bathing trunks. His suspicions turned out to be correct. They belonged to Noel Langsford. They were all that remained of the 22-year-old lifesaver.

In the gathering dark of the previous Wednesday the 187 centimetres tall Langsford disappeared without trace in the fold of a wave that suddenly turned red. The cry had been 'one last swim!' Four people plunged in. The water was refreshing, the surf thunderously good. They lingered. It grew darker. Ern Cross, 15, saw what happened from the beach. 'They were about to get a shoot when Noel dived under a wave and the shark got him. I saw him dragged through a wave, then I saw him disappear through another wave.' Those in the surf did not see the shark. Jan Faithorne was 4.5 metres from Noel when she heard him scream. She turned and he was no longer there. 'I saw a great gush of blood come from the wave and spread across the surface.'

Sixty people, including members of the Wamberal Surf Club, made a determined effort to find him. With car headlights lighting up the white flecks of foam, a variety of craft cruised back and forth till dawn without success. They were equally determined to get the guilty shark. A baited drum was anchored overnight.

Next morning the drum, dented in several places and scratched with large toothmarks, rolled up on the sand. According to the local constable, the strip of coast had been teeming with sharks but 50 gallons of bullock's blood tipped in to the sea failed to catch even one.

Newcastle

Newcastle is the second largest population centre in NSW. Like Sydney, 150 kilometres to the south, it is fringed by some of the loveliest beaches in the world. It is an industrial city and the sands and ocean are a magnificent contrast to the 'dark satanic mills' of the steelworks. Newcastle has seen some particularly savage attacks through the years. Some of these are now forgotten. Others still haunt the older generation.

It took a long time to forget what happened to 16-year-old Ron Johnson on 12 February 1948 because he was out there, with nine others, 45 metres from the shore at Stockton Beach when the shark alarm sounded. The anxious surfers bodied onto the next wave. Ron was not quite so lucky. He missed it. It crashed over his head and he remained behind while the others went careering into the shallows. It was a classic setting for an attack. And as Ron Johnson floundered towards the beach he was seized by the shark. Lifesaver Albert Linich splashed to the rescue with belt and line. But Ron Johnson, unconscious with shock and loss of blood from a mutilated leg, died in the ambulance room.

A shark surfaced in the shallows soon after. Incensed lifesavers pelted it with anything they could lay their hands on. A soldier, on the beach at the time, fired a shot at it and it disappeared.

Bar Beach is a magnificent sweep of sand and a great favourite among the people of Newcastle. Hundreds of thousands of its citizens have enjoyed its soft sands and exhilarating surf over the years. But their pleasure was marred on three separate occasions, in the 1920s, in the 30s and in the 40s.

Arthur Edward Lane, a 28-year-old Englishman of Janet Street, Merewether entered the surf, 135 metres south of the Cook's Hill Surf Club at 5.15 pm on the fourth day of April, 1928. He paused to make a lighthearted exchange with two girls standing waist deep together close to the water's edge. Lane strode on as far as the second breaker. They heard a cry and saw him lifted onto a wave, its white crest already washed with red. One of the girls splashed ashore to fetch help, while the other girl, Lucy

Donaldson, holidaying in Newcastle from Brisbane, went to assist the wounded man now being carried by the waves into the shore. His right hand was severed at the wrist, his right leg badly mauled. Lucy Donaldson was later commended for her aid. Arthur Lane died on the way to Newcastle Hospital.

It was 'good shark weather' one afternoon in February 1937, still, sultry, overcast. John Welsh, 32, of Denison Street, Hamilton, secretary of the Cook's Hill Surf Lifesaving Club, joined a dozen other bathers at Bar Beach at 3.20 pm. He was in 5 metres of water. There were others further out as well on either side. One of the bathers felt a swirl beneath him. A moment later a shark broke the water and seemed to pounce on Welsh, dragging him under.

Lifesavers had to smash open the door of the club casualty room. John Welsh had the keys. Five minutes after the attack his bloodless, mutilated body was washed ashore. It is believed the shark swam beneath the other bathers as it singled out its victim. John Welsh supported his widowed mother. The Cook's Hill club opened a subscription to pay for his funeral expenses.

The next attack at Bar Beach occurred twelve years later, 23 January 1949. It caused a gasp of horror throughout Australia, because it took place in the midst of a surf carnival before the eyes of thousands of spectators — including both parents of the victim.

Ray Land was a clerk in a small goods firm in Newcastle. His greatest pleasure was to get down to the surf. He was already Northern District Senior Surf Champion. That Sunday afternoon Ray was feeling tired and a bit chilled. However, he decided at the last minute to enter the contest for the sake of the team.

Ray was in 2 metres of water taking part in a heat of the district Rescue and Resuscitation. But the splashing and turmoil of a surf championship did not deter the 3.6 metre shark singling out the 20-year-old champion. It was seen at the moment of attack. 'The shark flicked its tail as though shaking its head as it took Ray down.' The surf boat gave the raised oar signal and in a moment the shark alarm went off.

In a flash, Reg Trew of the Cook's Hill Surf Club came paddling across on his surf ski. Using his feet and fists Ray Land had apparently driven the brute off. His hands were badly torn when Trew lifted him on to his surf ski and they both tipped off. He held Ray on the ski with his body while he pushed it along in the water until the surf boat reached them. Trew tried to staunch the flow of blood. As Ray Land was carried to the beach a dramatic call went out over the loudspeakers for blood donors.

Ray Land's distraught parents followed the ambulance to Newcastle Hospital. Behind them came a truck carrying 12 club members. Ray Land died 15 minutes after arriving at the hospital. Reg Trew's action was placed before the meritorious awards committee of the Surf Lifesaving Association. 'Trew acted up to the highest traditions of the association. His presence of mind and his courage in holding Ray Land on the ski while half in the water himself, the shark still near, were outstanding.'

The tragedy speeded up the meshing of Newcastle beaches which had lagged so far behind those in Sydney. Meshing, however, did not prevent an attack on Frank Okulich, 21, local surf ski champion, at Merewether Beach, Newcastle, on 6 December 1951. The exception proved the rule. Okulich was attacked on the one section of beach which was not meshed because of a reef.

He was surfing about 90 metres from the sands with some of his mates. A shoot came rolling in; Frank and Jim Jones, 15, missed it. The other pair sailed off. Jones began swimming towards the beach; Frank Okulich lingered for the next big wave. People on the beach said he abruptly disappeared beneath the surface and reappeared waving his arm. The wave he had waited for came. They witnessed a rare and bizarre frieze of a shark attack. The afternoon sun illuminated the wave. Through the shimmering water they glimpsed a silhouette of the man and beside him a long, dark shadow. His fatal injuries bore witness to the shark's savagery.

North Coast

Byron Bay, about 650 kilometres north of Sydney, is a haven for surfers. It was here that Thomas MacDonald, 16, was attacked on 23 October 1937. The shark first went for a youth on a surf ski but he drove it off with his paddle. It made for a second surfer who discouraged it by kicking furiously. Finally it turned upon MacDonald. It attacked him twice while he was swimming near a disused jetty. Fortunately a wave carried him to the shore before the shark came in for the kill. Although lacerated, MacDonald was stitched up and recovered.

The mental scars left on those related to the shark victim can be as abiding as those inflicted by physical injury. Time will never erase the memory of Stanley Elford. One minute the four Elford brothers were swimming together. In the next anguished minute 22-year-old Stanley Elford was carrying home the mutilated body of his dead brother, Edwin, in his arms.

The boys had swum together many times in the brackish waters of the Hastings tributary, close to their home. The spot was 20 kilometres inland from Port Macquarie, a popular holiday resort on the NSW north coast, 430 kilometres north of Sydney.

It was four in the afternoon on 8 November 1947. Rupert, 13, screamed and began kicking frantically. His knee was badly cut, either from the shark's teeth or its fin. The shark darted to the next brother, Edwin, 12, and seized him by the knee. Stanley, the eldest, was there in a flash. He tried to tug Edwin away from the shark. Suddenly he succeeded. The shark let go and shot away. It took the boy's leg with it. Edwin died as his brother brought him to the bank. Meantime, Rupert, with severe cuts to both thighs, stumbled the 50 metres to their home. He recovered later in Port Macquarie Hospital.

An act of heroism by surfer Paul Howard in February 1977 cost him wounds to his left hand, arm and leg and a large hole in his surfboard. Rain was falling at Kingscliffe Beach, south of Tweed Heads, but about 20 or 30 surfers were in the water. Paul, 24, saw a 4 metre shark close to a young girl on a rubber surf mat. He called out and moved close to the girl, punching her surf mat out of the shark's path. The shark turned on him, biting a large chunk out of his board and continuing to attack Paul and the board as he paddled to shore.

The young hero coolly walked up the beach, refusing ambulance transport. 'Get that beach closed,' he told authorities. However, he was probably more affected by his ordeal than he thought. As he drove to Tweed Heads Hospital he was involved in a car crash. Although he escaped further injury, 80 stitches were inserted in his wounds.

On 7 March 1982 20-year-old Marty Ford was attacked by a shark while surfing with five friends at Tallow Beach, south of Cape Byron. Marty was 200 metres from the beach when two of his friends saw him go under and come up screaming for help.

Somehow they pulled him back on to his surfboard. His legs had been badly mauled. They carried him back to the beach where they applied a tourniquet to one of his legs. An ambulance arrived. By the time it reached Byron Bay Hospital Marty Ford was unconscious. He died soon afterwards.

This had been the first death by shark attack in NSW for 19 years, since the fatal attack on Marcia Hathaway in January 1963 in Sydney's Middle Harbour. By a gruesome coincidence the next shark fatality in NSW was to come 11 years later when another Ford was taken, also at Byron Bay.

The belief that you are safe from shark attack when dolphins are around is a myth. Yet Adam McGuire probably owes his life to a school of dolphins which chased off a shark which attacked him at Half Tide Beach near Ballina, south of Byron Bay, on 3 January 1989. Adam, 17, and two of his mates had travelled from Newcastle on a holiday to the North Coast. At about 5 pm the boys were riding 1.5 metre waves among a school of 15 to 20 dolphins, when they noticed the dolphins becoming restless.

'I looked over and saw Adam knocked off his board,' said Brad Thompson. 'I could see a hole in his board, and Adam in the water close to it. I saw the shark come up to him. He started hitting it on the head, trying to get it away from him. Then we didn't see it any more and we took Adam in to shore.'

Adam was treated for a severely lacerated abdomen. From an examination of the bite mark on his surfboard the shark was tentatively identified as a 4 metre tiger.

SYDNEY RIVER ATTACKS

The earliest recorded river attack occurred on the Macleay, which flows into the sea at Trial Bay, 386 kilometres north of Sydney. The man, Alfred Australia Howe, was taken on 17 January 1837. He was the grandson of the first Government Printer and the son of the owner of the *Sydney Gazette*, Robert Howe, who died in a boating accident on Sydney Harbour eight years earlier, when he tried to save young Alfred from drowning.

Lane Cove River

The first river attack to arouse public concern took place on Friday

26 January 1912 on the Lane Cove River near Fig Tree. James Edward Morgan, 21, was swimming with two friends. He cried out, 'Help! A shark has got me!' The shark was seen tugging his struggling body through the water.

Two days later a 'grey nurse' was caught close to the spot where he was taken. Shark authority David Stead (the father of author Christina Stead) had insisted the shark was a whaler and he was right.

The story had a bizarre touch. The 2.8 metre long shark was put on exhibition to the public for a fee. The proceeds were given to Jim Morgan's widowed mother. Later when the shark was opened up they knew it was the right one. Human flesh, analysed as belonging to the dead man, was found in its stomach.

Parramatta River

Sharks have been sighted in Sydney's Parramatta River on many occasions. On Christmas Day 1930 Jim Knight, swimming in the Homebush Bay area, was fortunate in receiving nothing more than several bumps from a shark, resulting in abrasions and a thoroughly bad scare.

A fatal attack did occur on the Parramatta River — around 70 years ago, at 4 o'clock on a hot Saturday afternoon. Today, no one would dream of swimming in the polluted waters running through heavily-industrialised Camellia. In January 1924 it lived up somewhat to its pretty name. With a mere handful of factories, the water was still reasonably untainted and several youths were cooling off in its shallows. Out of nowhere a 3 metre shark made a savage attack on 16-year-old Charles Brown. His mate, Fred Cox, 17, tried to save him, but in minutes, Charles was dead, his body badly lacerated.

On 18 January 1960 a 3.5 metre shark approached a rowing shell in the river, near Concord Repatriation Hospital. It began cruising alongside and the four rowers kept on rowing. The shark kept company with them for 180 metres. On the previous day, Sunday, the crew of a pairs shell on the Parramatta River had tipped over into the water. Instantly a dorsal fin appeared close by. The two men climbed back into their shell in a flash and rowed off with a speed that would have broken all world records.

Georges River

On two occasions swimming out to retrieve a tennis ball has invited shark attack. It happened at Ellis Beach near Cairns in 1946, as it happened years before on the Georges River at East Hills, in January 1934, a year darkened by the shadow of the shark.

Frank Spruce was sitting on the steps of Lambert Wharf, watching the crowd splashing and swimming in the water, most of them outside of the nearby safety enclosure. One young man struck out from the body of the crowd in hot pursuit of a tennis ball which began drifting midstream. A fisherman on the wharf said later, 'I saw a dark shadow dart in his direction. It was all so quick. I saw the boy lifted clean out of the water, the shark's jaws on his chest and arm.'

Frank Spruce saw it too, a moment before the attack. He shouted a warning and simultaneously plunged into the water. He grabbed the 15-year-old boy, Wallace McCutcheon, and dragged him to safety. His prompt action might well have put the shark off from coming in for a second attack.

Picnickers crowded around the wounded boy who remained conscious. Many of them patted Spruce on the back. But there was something worrying the boy more than his injuries. 'Don't tell Mum!' She was ill at the time and the news may have been too much for her. He was taken to Canterbury Hospital with wounds to the chest and back. 'Another half inch,' said Dr Abramovitch, 'would have pierced his heart.'

At Deepwater, Hollywood and Kentucky, along the George's River several dogs had been taken by sharks in previous weeks.

Richard Soden, 19, living in Canberra, was spending Christmas with his foster parents at Moorebank, Sydney. Their home was 800 metres from the Georges River. Late in the afternoon of the last day of 1934, Soden and two other young men were swimming in the river near Milperra Bridge, a popular bathing spot more than 30 kilometres from the river's mouth. At that point the distance between banks is about 45 metres and at 4.30 the lads decided to race each other across.

A fisherman standing on the bank watched idly as Richard Soden took the lead. The man suddenly froze with horror at the sight of a dark fin cleaving the water behind them. For a moment he lost his voice. The word came, loud and hoarse, 'Shar-ark!'

The dorsal disappeared. Richard Soden, now 20 metres from the opposite bank, let out a scream. A huge tail momentarily thrashed the water and the youth disappeared from view. A long minute later he reappeared swimming very slowly. Two fishermen jumped into the water and waded out to the channel to meet him with outstretched hands.

Severe shock and a badly mutilated leg had taken their toll. Richard Soden finally did make the opposite bank. But he was dead when he got there. His foster brother, who had been in the race, hysterical with grief, refused to believe there was a shark, and insisted Richard was still alive.

But several fishermen knew a shark had been around. Early that morning George Markham and his mate were puzzled by the shoals of passing fish. 'There's your reason!' The two men watched as a large shark cruised lazily by in the shallows a few metres from the bank.

It is unlikely that news of the tragedy at Milperra Bridge reached the ears of Mrs Morrin of Bankstown later that evening when she took her five young children for a swim at a picnic spot called Kentucky, 5 kilometres further up river. They were accompanied by a friend, Jim Schofield.

At the spot where the youngsters entered the water, the river is 90 metres across; it shelves gradually out to the channel in the centre. While Mrs Morrin chatted on the bank with Schofield the children splashed and laughed together. There was no thought of danger. They were no more than chest-deep, about 9 metres from the bank. Lots of families came here to swim, although there were none around on that warm, moonlit night, less than four hours away from 1935. Besides they were 38 kilometres from the river mouth and no one had ever heard of a shark there, not even older residents.

Beryl Morrin, 13, and her brother Tom, aged 10, challenged each other to swim to a sandbank, a few yards further out. The attention of those on the bank was attracted by sounds of loud splashing out in the inky waters. Then Beryl Morrin uttered a piercing scream. The ghastly scene that confronted Mrs Morrin and Jim Schofield was mercifully shrouded in the semi-darkness. Beryl Morrin stood in the water holding up two stumps where her arms had been. Jim Schofield flew into the water and seized the sobbing child above the elbows squeezing them to staunch the flow of blood. The child's mother was right behind him.

Australia's worst year on record for shark attacks had ended. People opening their newspapers next morning were aghast at the double tragedy, only three hours and 5 kilometres apart. Dr Coppleson's theory of the 'rogue shark' seemed to have been given some credence. The wave of revulsion was sharp-edged by the desire for revenge. A net was strung across the Georges River at Kentucky and shark fishermen were out in force. Bill Kelso hooked a 3.6 metre shark but the hook snapped under the tremendous strain.

At Milperra, on 1 January 1935 a notice was nailed up:-

DANGER — SHARKS
Please Use Baths

In spite of the warm weather and the holiday, only small numbers ventured into the surf along Sydney's beaches that day. Meantime at Canterbury Hospital doctors fought for Beryl's life. A Beryl Morrin Fund was opened by the Mayor of Bankstown and contributions poured into the Sydney newspapers. On Thursday, three days after the attacks, the funeral of Richard Soden took place. For Beryl it was still touch and go. The 13-year-old girl had youth on her side. There was a universal sigh of relief with the news that Beryl Morrin had rallied. A few days later it was reported she was out of danger. Even the most awful twist of fate did not always end in total disaster.

More than one rescuer has lost all sense of time in a frantic, face-to-face confrontation with a killer shark. To Rudolph Tegal it seemed 'like an eternity' as he held off the creature from his teenage daughter, cradled in his right arm. 'I thought the shark would go away at any moment.' In the melee in Middle Harbour, 17 years later, Frederick Knight also lost all sense of time as he fought off the shark attacking Marcia Hathaway.

On Saturday 7 January 1946 Mr Tegal and his daughter Valma, 14, and his son Owen, of Oatley, together with Barbara Walther of Villawood, were swimming in Oatley Bay on the Georges River. Young Owen decided to pack it in after a while, but his sister stayed. She dived off the jetty, swam close to the bank and rose to her feet in a metre of water.

She cried 'Dad, quick!' The simple words that followed almost blurred their tragic meaning. 'My leg's gone!'

'I grabbed her with my right arm,' her grief-stricken father recalled. 'I found the shark, big monster, between my legs ... I lashed out with the other arm at the shark, calling to my son Owen for help.' The shark may have taken fright. It did not attack a second time. It had done its work. Valma, her left leg missing, died soon after being brought to the bank.

At the Australian Museum, G.P. Whitley issued a statement: 'The only safe precaution against shark attack is to bathe in the protected areas. Attacks cannot be anticipated, and their likelihood does not vary with the time of the year or locality.'

CHAPTER NINE

Queensland

Waterside worker, Herbert Jack, 40, and his mate Thommo, both of New Farm, decided to take both their young sons for a fishing expedition on a sunbaked Brisbane Sunday morning, 27 November 1921. Their trip ended in injury and heart-break before it had even got started.

Herb Jack hoisted his 8-year-old son George on to his shoulders and began to wade out to a small dinghy a mere 9 metres from the bank at a spot known as Gay's Corner along the Bulimba Reach of the Brisbane River. The man and his son were in water a mere metre deep when the incredible happened. A shark seized him by the right hip. He managed to hold his balance, gripping his son's legs with one hand as he lashed out at the fish with his right arm. Its teeth tore a nasty gash.

Thompson, who was watching unbelievingly from the bank, jumped into the water. In the melee that followed, young George slipped from his injured father's shoulders — or he was tugged off by the shark. The two panic-stricken men stared about them. The shark had gone and so had the boy. Then, for a split second, he surfaced several yards away. The two men, one supporting the other, staggered towards the spot. But he had gone. Moments later a police launch happened to be passing. They called and waved. Sergeant Henderson organised an immediate search. Herbert Jack was taken at once to Brisbane General Hospital. He recovered to face the lasting tragedy of his lost son.

That attack strengthened Brisbane people's caution toward their river. It accounts for the passing of some forty years before another reported attack. It happened at 9 o'clock on a December morning in 1960. Fortunately the shark was a 'littley,' a metre long, and the man, Lester McDougall, received minor injuries.

Considering the waters along the immense Queensland

coast are warmer for longer periods of the year, shark fatalities have been few and far between. There were only six recorded attacks throughout the whole of the 1920s; seven in the 1930s.

One of the most shocking tragedies occurred away from the densely populated areas. This was the double fatality at Kirra Beach, Coolangatta on 27 October 1937. Four days previously at Byron Bay, 50 kilometres south of Kirra, Thomas MacDonald, 16, had been attacked (see Chapter 8).

Coolangatta, October 1937. The Doniger brothers pose beside the 2.6m (11 feet 9 inch) tiger that took Norman Girvan and Jack Brinkley at Kirra Beach the previous day. Human limbs were found undigested in the shark's stomach.

Norman Girvan, 18, and Jack Brinkley, 25, were savagely mauled in the space of minutes. One died there and then. The other, following amputation, died in hospital. The third man, Gordon Doniger, who left the sandbank to swim the 183 metres to the beach, gave this dramatic story to the local police.

'At about 5.15 pm on 27 October 1937, Norman Girvan and I entered the water at Kirra together. I swam out until I was about 150 yards (135 metres) from the shore. I was then on the sand spit. I waited there until I got a shoot, and came in on it over the spit to the gutter which is about 80 or 90 yards from the shore. When I got there I met Girvan who had just swum out. We passed some remarks, in fun, about sharks, and both of us tried to see if we could touch the bottom, which I did and found to be about 20 feet (6 metres) deep.

'I saw a shoot coming and called out to Girvan, "Here's a shoot!" I swam to the other side of him but about 5 yards farther out. Just then Girvan said, "Quick, Don, a shark's got me!" I thought he was still fooling and called out to him not to miss the shoot. Just then he put his arm up and I saw blood shooting everywhere. I could tell then that he had been attacked by a shark, and I called out, "I'm coming, Blue." (Girvan is known as "Blue") I swam over to him and took hold of him and at the same time I saw a man named Jack Brinkley swimming towards us about 10 yards away. I called out to him, "For God's sake, Jack, come and give us a hand with Blue and get this thing away".'

'At that time I had hold of Girvan. I first tried to take him by the arm and found that it was just hanging by a bit of flesh. Brinkley turned to swim towards us and just then he began to kick and struggle and it appeared as though he had been attacked. I was trying to get to the shore with Girvan at this time. Girvan said to me, "It won't let me go, it's got my leg." Just then I felt Girvan being shaken forcibly and he was pulled out of my arms. I felt the body of a shark brush my thigh. The shark came to the top and I could see that it was a big one, but I cannot estimate its length. Girvan said, "I'm gone. Goodbye," and almost immediately the shark dragged him under the water. I then struck out for the shore. I was swimming with my head down in the water when a shark swam right underneath me. I thought it was going to attack me and I tried to push it away and my hands caught it by the fin. I pushed at it and pushed myself away and I saw the shark attack Brinkley at his left side, and just then I got on a shoot and came into the shore. While on the shoot I passed my brother who was swimming out to Brinkley. I remember arriving at the shore but I do not remember anything after that for about 10 minutes.'

Joseph Doniger continues:

'I was standing about ankle-deep in the surf at Kirra, Coolangatta. I saw three men, Jack Brinkley, Norman Girvan and my brother, Gordon Doniger, about 75 to 80 yards from the beach. I saw the three men waving their arms about and splashing in the water. I heard Brinkley yell my name and something about a shark. I then saw Brinkley begin to swim towards the shore. I heard my brother Gordon calling to Brinkley for assistance. I saw a swirl in the water become discoloured and red with blood. By this time Brinkley had turned and was swimming back towards my brother and Girvan.

'By this time I had entered the surf and was swimming towards the three men. As I was swimming out I saw Brinkley, who had only then taken a few strokes towards my brother and Girvan, attacked by a second shark. I know that there were two sharks because the shark that I first saw was bigger then the shark which attacked Brinkley. By the time Brinkley was attacked, Girvan had disappeared and my brother was swimming towards the shore. I was then within about 5 yards of Brinkley who was calling out to me, "Hurry up, Joe, for God's sake get me out of this."

'I swam to Brinkley and secured a hold and began to make for the shore. Just as I did the shark attacked again. This time it caught Brinkley by the left arm just below the shoulder. Brinkley gave a groan, stiffened out and became a dead weight. His left arm was bleeding freely and was practically severed.

'When the shark attacked Brinkley when I was holding him I saw it plainly, it was about 8 feet long. The shark made no further attacks and I then succeeded in getting Brinkley to the beach where, with the help of other members of the Kirra lifesavers, a tourniquet was applied to Brinkley's arm. Brinkley remained conscious during the whole time I was with him. All I heard him say was, "If I hadn't turned back for Blue it wouldn't have got me." I did not see Girvan after he sank out of sight from my brother's arms.'

Norman Girvan was never seen again, not in one piece. Parts of his body drifted inshore later. When a 3.6 metre tiger shark was caught following a grim hunt the next day, portions of human limbs were found undigested in its stomach. Among them was Norman Girvan's right hand, recognisable by a scar.

Everything was done at Coolangatta Hospital to save Jack Brinkley. It was a lost cause. He died shortly after his left arm was amputated.

Could it have been the same shark that attacked Tom MacDonald four days earlier? Sharks are known to swim up to 48 kilometres and more per hour. In that amount of time it could have moved between Byron Bay and Kirra Beach.

Of the nine recorded attacks in the 1940s, six were centred in the Cairns area. The attacks, the first ever known along those beaches, began with the mauling of two wartime servicemen.

The first was Flying Officer, Athol Wearne, RAAF age 24, who lost a leg while swimming at Trinity Beach, 17 kilometres north of the city, on 12 September 1942. Someone shouted a warning but the shark grabbed him before he could make the beach. A doctor who was with the party gave first aid and Wearne eventually recovered.

Three years later a second Australian serviceman was savaged and died off the same beach on 15 June 1945. He was only a few yards from the beach at the time. His two mates got to him at once, but it was already too late.

The third attack at Trinity, ten months later, on 19 April 1946 was also fatal. Bob McAuliffe, 17, was hip-deep in water less than a man's length from the sand when a bather saw a 'brown mass' move in and take him. McAuliffe died of lacerations to the legs and hands before he reached the hospital.

The scene of the next attack in the Cairns area shifted to Ellis Beach, 30 kilometres north of the city. On 18 August 1946 Phillip Collin was with a picnic party. At about 3.20 pm they were having fun with a tennis ball when someone missed their catch. It was a fatal miss. The ball hit the water and started to drift out to sea. Collin went after it. The ball, drifting just ahead of him all the time, led him into the jaws of a shark. His horrified companions witnessed the attack. His body was never recovered.

A party made the two hour journey from Tolga in the Atherton Tablelands for a Sunday at the beach. Included were the local postmaster, Mr Maguire, his wife, and Ted Stockwell, Tolga's police constable. Richard Maguire travelled up from Cairns to join his parents at Ellis Beach. He was a boarder at the Marist Brothers' College.

The hour of 1.30 pm 17 April 1949 would stand still for all time for the Maguires. In water, hip-deep, Richard Maguire was savagely mauled by a tiger shark. It was over in seconds. There was a cry of 'Shark!' Ted Stockwell, a dozen feet away, flung himself to the boy's side, as Richard's father dashed in, fully clothed. The shark vanished. Richard's leg was severed at the hip. He died almost at once.

The Cairns Surf Lifesaving Club set baited lines to floating petrol drums. A shark about the size of the one which attacked Richard Maguire, was caught. There was no proof it was the killer. It would have been small consolation if it had have been.

The last attack in the area came 18 weeks later, at the same time of the day, at Yorkey's Knob Beach, 9.5 kilometres below Ellis Beach. Jim Howard, 34, was swimming about 130 metres from the shore when a shark crunched his leg. Somehow he managed to wrench free and was rescued by Brian Ware, 21, holidaying from the Sydney suburb of Alexandria. Howard died on the way to hospital. Brian Ware was also taken in the ambulance. At first it appeared to be a rare instance of a rescuer being attacked in turn. The shark had merely brushed against him and its abrasive body caused nasty injuries to his stomach. Was the same shark responsible for all the attacks?

Almost 1,600 kilometres north of Brisbane lies the tropical city of Townsville. The giant meatworks where North Queensland cattle is prepared for export is located on the Ross River, a few miles out of the city. With so much offal discharged into the river, the 'scavengers of the sea' were there in their swarms to clean up. There were five attacks spread over 18 years. There should not have been any.

Jack Hoey, 38, went for his last swim there, on 5 January 1919. Robert Milroy, 54, was swimming in the river in 1922. Several sharks tore him to pieces. Edward Hobbs, 42, died on 1 September 1929. It was pure accident. He slipped and splashed into the creek from a wharf. A shark struck the moment he hit the water and he died instantly. Eighteen months later, a teenage Japanese-Australian, Arthur Tomida, 19, was savaged to death. His shrimping net got tangled and he waded out to fix it.

The death of Bill Tennant, 33, on 16 May 1937 might well haunt the dreams of a Ross River ferryman. The ferryman would not allow Tennant to board the ferry. It was crowded and he would have to wait for the next trip. Tennant decided to go anyway. He dived into the river and began to swim across. He was only 10 metres from the bank when he was attacked. Some of the passengers witnessed the attack and the ferry turned back. The shark took both his leg and arm and he died before he reached hospital.

A somewhat bizarre tragedy that probably had to happen in Australia sooner or later, occurred at the Kissing Point Baths in Townsville, 2 October 1951. In company with a neighbour and his family, Albert Kenealey, 42, a waterside worker of South Townsville, had gone to the baths for a safe evening's swim at the close of a hot day. Kenealey was the last person left in the pool and was about to come out.

There was a sudden rush of water and Kenealey shouted 'Eddie' — the first name of his neighbour — and disappeared. Eighteen metres outside the enclosure, in Cleveland Bay, they found the body of Kenealey, shockingly mutilated with one leg completely gone.

Senior Sergeant Harley of the Townsville Police said the wire surrounding the baths had corroded. The baths belong to the Country Women's Association who had recently appealed for help for improvements to the baths. The delay cost Albert Kenealey his life in a place where he believed he was safe.

On 2 November 1975 the Sydney Sun-Herald published a photograph of a gaping hole 'big enough for two men to swim through' that had rotted in the 50 metres 'protective' shark net across Parsley Bay, Vaucluse, a spot popular with young children on Sydney Harbour.

Shortly after midday on Sunday 4 September 1955 a radio message was received at Rockhampton, Queensland, from the coastal freighter *Caledon*: 45 FOOT KETCH 'CHRISTINE' DRIFTING IN A DANGEROUS POSITION HERE. MASTER PROBABLY INJURED. The news sent Harbour officials and detectives 35 kilometres down the Fitzroy River to a spot known as Humbug Heads.

When they arrived the ketch was aground. As the men squelched across the mud they saw an arm hanging out over the side of the tilted vessel. The sight on the deck made even hardened detectives gasp. A man was lying amidships, one arm dragging in the mud, the other tangled in the ship's rigging. His right leg was torn off at the hip. He was disembowelled by a gaping stomach wound. The ship's rigging had prevented the body from slipping over the side.

The ketch had sophisticated communication equipment including two-way radio and an 800 kilometre range radio telephone. It was quickly ascertained that the dead man was Dr Ernest Joske, 56, of Adelaide who was sailing from Brisbane 1,360 kilometres north to Bowen, one stage of a round-the-world trip. Several spanners lying around the deck suggested Dr Joske may have had engine trouble. He had radioed the government radio station at Rockhampton on Friday, stating his position. On the afternoon of the following day, fishermen had spotted the ketch near the mouth of the Fitzroy River. It was under sail and appeared to be in perfect control. Joske had not been dead more than 12 hours.

Police eventually decided one of two things had happened:

he was attacked by a shark when he had gone overboard to fix the propeller and had somehow managed to heave himself back on deck before dying. Or he was mauled while leaning over the side. No one will ever know the precise answer to the riddle of the mutilated doctor on the deck of the *Christine*.

It was the first time it had ever happened at Queensland's favourite holiday resort, Surfers Paradise, and the local lifesavers responded in the finest traditions of their organisation. The beach was crowded on Sunday 23 November 1958 and Peter Spronk, 21, was doing all the right things. He was swimming between the safety flags, not too far out, about 20 metres from the beach beyond the second line of breakers. Eagle-eyed lifesavers on the sands spotted the attack almost the moment it happened.

'They responded to the alarm en masse,' club captain Claude Jeaneret, said proudly afterwards. 'Several of the lifesavers entered the water without a moment's hesitation.' The shark cruised close by as the lifesavers brought the dying victim out. Two of the beltmen, Eddie Burnett, 17, and Peter Brennan, 21, collapsed with exhaustion when they got back to the beach.

Queenslanders were stunned by two appalling shark attacks that occurred in the waters of their state in December 1961. The Andrews family, from the Brisbane suburb of Sherwood, were staying at a holiday cottage at Noosa Heads, a scenic resort about 240 kilometres north of Brisbane. The killer shark's sudden presence out of nowhere, on a beautiful stretch of beach brings to mind the appearance of the serpent in the Garden of Eden.

Twenty-two-year-old John Andrews, a dental student, was up really early with his surf board on the morning of 18 December 1961. While his parents and most of the other holidaymakers were fast asleep in their beds, John was enjoying the exhilarating experience of flying beachward on the curl of a wave. At 6 am he decided to go back for some breakfast. He was in 0.6 metres of water, pushing his board alongside him, a couple of body-lengths from the sand, when the unbelievable happened.

John Andrews screamed out when the 3 metre shark attacked him. Retired planter, 68-year-old Rawdon Payne was taking his morning constitutional when he heard it. 'The water was so shallow the shark's whole body was exposed,' he said later. Payne rushed in to help. 'I started to pull him out. The shark kept trying to get at him.'

People came hurrying over the sands still in their night attire. They included John's parents. He lay on the beach, his left leg and his right arm almost severed from his body. Bravely his mother knelt beside him, spooning tea into his mouth and whispering comfort until medical help arrived. At Nambour Hospital, John Andrews was given transfusions and minor surgery. Highway police cleared the way for the 100 kilometre dash to Brisbane. John died the following Sunday.

At 4.15 on Thursday afternoon, 28 December 1961, four days after John Andrews died in Brisbane Central Hospital, 1,160 kilometres north a laughing group of young people were sitting on Lamberts Beach, near Mackay. Margaret Hobbs, 18, a student schoolteacher and her boyfriend, 24-year-old Brisbane salesman, Martin Steffens, decided to go and wash the sand off in the sea. They waded out, hand-in-hand and in waist deep water, about 5 metres from the edge, Margaret playfully began splashing Mark. Laughing, he caught her up in his arms and began dunking her. Her screams of laughter became a scream of terror. Margaret had been snatched from his arms by a large shark who had her thigh in its jaws. Martin Steffens lashed out at it with all his strength, his hand already in a mess.

It took some moments for those on the beach to realise the young couple were no longer having fun. There was a long scream and a swirl. The shark made three rushing attacks at the couple. Without hesitating, Graham Jorgensen, a 27-year-old engineer, plunged out to the stricken pair.

With her right arm gone and her left hand missing, Margaret Hobbs incredibly remained conscious on the beach while a nurse who was on the beach applied tourniquet pads. An ambulance took them to the Mater Hospital at Mackay.

Margaret proved to be a rare type A positive, and people flocked to the Red Cross blood bank that evening in answer to a radio call for donors. Nothing could save the young schoolteacher. Mercifully, perhaps, she died. Mark Steffens survived, the mental scars never to be entirely erased.

There could be no more shillyshallying. The Queensland Government was forced to act. In consultation with Nick Gorshenin, who had been contract-meshing Sydney's beaches from 1948 to 1962, Mr Nicklin, the Premier, at last approved the first meshing for Queensland's beaches. On Saturday 3 November

1962, nets were laid by the Harrises on the south coast beaches from Southport to Coolangatta and on the north coast beaches from Caloundra to Noosa.

The meshing of Queensland's beaches, however, did not provide any protection for surfers or fishermen at Stradbroke Island, 30 kilometres from Brisbane. Bruce Lawler, 16, was paddling his board in a surfing competition off Stradbroke on Monday, 27 August 1973 when a shark jumped up and grabbed his foot. He managed to hang onto his board as he was dragged down twice, throwing punches at the shark, but to no avail. 'It was pretty big ... I opened my eyes under water and saw it — I could see the grey.' Bruce lost his right leg to his attacker.

On Saturday night, 31 August 1974 Peter Hodgson, 31, and his next door neighbour, Adrian Treveluwe, 30, took off in their 3.3 metre boat for a fishing expedition off Stradbroke Island. At seven the following morning they decided to pull up anchor. Treveluwe started the boat but rode over the anchor. When he took up the slack, the boat whipped around, and water poured into the stern. In minutes the boat went under. The two men managed to grab their life jackets. Their fishing trip was to end in nightmare and death.

The great white shark

They drifted with the current but had to kick and swim when the waves rolled over their heads. Then they spotted the black fin moving towards them. Adrian Treveluwe was nearest in line. As the fin came up, Hodgson shouted, 'Kick at it!' The man replied with a stifled groan and went under. Peter Hodgson grabbed his friend and kicked against the dark bulk of the shark. Mercifully it went away. Hodgson held on to Adrian Treveluwe's torn body, expecting the blood to bring the shark or, more terrifying, several sharks, around at any moment. Ten minutes later the wounded man went limp and he died. The body drifted away.

Peter Hodgson drifted helplessly all through that Sunday and all through the long night that followed. After a time sheer physical and mental exhaustion lulls the senses, quietens the fears. Dawn broke. 'I could see the tall buildings of Surfers Paradise. One forgets the emotions, the moments of cold terror of shark attack.' When he was dragged from the shallows by one of the chefs from the Broadbeach Hotel, all Peter Hodgson could say was, 'Thank God. Thank God.'

Another fishing expedition off Stradbroke ended less tragically despite a close encounter with a shark in June 1976. Debbie McMillan was sitting in the front of her father's 5 metre boat when a 1.5 metre shark jumped aboard and hit her in the face, knocking her unconscious. 'I saw this thing come out of the water,' Debbie said. 'The next thing — bang, it hit me in the face. I blacked out and when I got up I didn't know what had happened.'

Debbie's father, Frank McMillan, said he thought the shark had jumped out of the water to try to shake off a sucker fish. The boat hit it in mid-air. Debbie's face was badly cut. The shark didn't get off so lightly. It was killed with a tomahawk.

Sometimes fate has a heavy hand. After surviving 33 hours in the water following a boating accident John Hayes, 45, and Victor Beaver, 74, were taken by sharks less than an hour before help, in the form of a passing boat, arrived.

With Verdon Harrison, 32, the two men were on a fishing expedition in a 9 metre launch on the night of Friday 11 March 1977 when a 25,000 tonne Japanese freighter, carrying a cargo of motor cars into Brisbane crashed into them near Moreton Island in Moreton Bay. No one on board the freighter noticed that they had hit the little boat as the three men struggled to find life jackets and free a dinghy from the boat. Unable to do either they clung to an ice-box which they found floating in the debris.

Throughout Friday night and all day Saturday they drifted up and down with the tide as they were gradually pulled further from the island. At about 4.30 am on Sunday sharks appeared. For 45 minutes they hung around before a shark grabbed hold of one of them. 'Goodbye, mates, this is it,' he managed to say, before he was dragged under the water. The second man maintained his grip on the ice-box for a while, despite being attacked by a shark. Soon afterwards, however, he too was dragged under.

Verdon Harrison was the next target for the sharks. He kicked at them as he continued to maintain his grip on the ice-box. Blood flowed from his feet. Then at 5.45 am help appeared. A charter fishing boat operator, Richard McMullin, spotted a blue object floating in the Pearl Channel. It was Verdon Harrison clinging to the ice-box. He collapsed in deep shock after being dragged onto the charter vessel. Tragically, the departure of the boat had been delayed half an hour because some of the party were late. Had the boat left on time it might have arrived in time to save the other two men. Later in the same year, on 24 August, George Walter died at Buddina Beach, south of Noosa, after his left arm was severed and his legs mauled.

Six years later two more people fell victim to sharks after an accident at sea further north on the Queensland coast. On the night of Sunday 25 July 1983 a fishing trawler capsized in heavy seas east of Townsville on the Great Barrier Reef. On board were the skipper, Ray Boundy, 33, a deckhand, Dennis Murphy, 24, known as Smurf, and Murphy's girlfriend, Lindy Horton, 21. The three managed to keep afloat using a lifesaving ring, three large pieces of foam rubber and a surfboard. Throughout the night and the following day they kept in good spirits, laughing and joking as they assured each other they would make it, despite seeing a number of sharks in the water.

At 7.30 on Monday night a shark, over 4 metres long, commenced circling them. The three remained unconcerned and the shark disappeared. Ray Boundy, who had lifted his legs out of the water, put one foot back down. The shark bit his left knee. 'I sort of panicked a bit and jammed my other foot down on him and he just let go,' Boundy said later. 'I said to Smurf, "We're not ready to be his dinner just yet." Five minutes later we were hit by a wave and knocked off our pieces of foam and as we came to the top it grabbed Dennis's leg and wouldn't let go. I yelled at him, "Kick as hard as you can." And he called out to me, "The bastard's got my frigging leg".'

After taking Murphy's leg the shark swam off but soon afterwards returned. Murphy, realising that with blood streaming from his body he was now a danger to his two companions, asked Boundy to take care of Lindy before swimming away. The two watched in horror as the shark picked up the top of his body and dived under. 'He was just screaming and I couldn't believe that anyone could have that much guts to get his mates out,' said Boundy. Murphy disappeared in a swirling mass of blood and flesh. 'It was like watching a human being fed through a mincer. Lindy was hysterical and I abused her and slapped her a couple of times and she came good.'

About 4 am the shark returned and circled. 'I saw it go for her,' Boundy recounted. 'It flung itself into the air and got the top half of her and turned her upside down. It was just so quick and she squealed and it shook her like a rag doll to get her out of the life-ring. I just had to bolt. She didn't say anything but was mumbling. She didn't know what had happened.'

Ray Boundy remained afloat for a few hours, trying to make it to the sanctuary of a nearby reef. Then he saw the shark again. 'I just couldn't believe it. I had lost the two best friends I ever had and he had come back for me. He followed me around for hours and I started to get worried.'

At last Boundy saw the reef. At the same time a rescue plane appeared. He managed to catch a wave over the outer reef. Safely inside he shouted out, 'You won't get me, you bastard.' He scrambled onto the reef, laughing and then collapsing in tears. Shortly afterwards an RAAF helicopter, alerted by the plane, arrived to pick him up from the reef and take him to Townsville Hospital where he was treated for the bite on his knee.

Ray Boundy was certain there was only one shark involved in the attacks. However, experts doubted this. According to Manly Marineland curator Chris Warner, it is extremely unusual for a shark to take two people, presumably eating them, and return for a third soon afterwards, even if it were very hungry. The attacks would most likely have been the work of a tiger shark or a reef whaler. Only a white pointer would be likely to take three people in such a short time and they are extremely rare in the warm northern Queensland waters. Ray Boundy was unimpressed by these opinions. 'I don't care what the rumours are or what the experts say,' he said. 'All I know is that I lost two of my best friends.'

The next shark victim in Queensland was sixteen-year-old Nicholas Bos, who was sailing on a catamaran with three friends a kilometre off Black's Beach on the central Queensland coast on Friday 31 November 1984, when he fell into the water. Nicholas was mauled by a tiger shark and died as the result of his wounds. Two days later a fisherman caught a 3.2 metre tiger in a bait net near a popular tourist resort at Airlie Beach, 50 kilometres away. It was impossible to say whether it was the same shark.

When a mayday call went out from a trawler returning to Yeppoon in northern Queensland on the night of Sunday 7 November 1988 several boats and aircraft went out in heavy seas and gale force winds in search of the fishing party. Shortly after sending out the call James Coucom, 33, his brother Bruce, 17, and his father Cedric, 62, left the sinking boat. James made it onto their 4 metre aluminium dinghy. His father and brother were not so lucky and he had to look on helplessly as they were eaten by sharks.

While eight aircraft, including two helicopters from Dutch navy ships, and a number of boats searched the seas, James Coucom drifted on the dinghy for two days without food or water while strong winds carried him over 150 kilometres. Late on Tuesday night a National Sea Safety Surveillance helicopter spotted him in the half-submerged dinghy north of Townshend Island, 130 kilometres north-east of Rockhampton. Suffering from exhaustion and dehydration and no doubt profoundly shocked by his tragic loss, James Coucom was taken to Rockhampton Base Hospital. He was still hallucinating when police came to interview him.

Shark stories are, of course, horrifying, but if sharks could speak they would probably have some horror stories to tell about humans. Many more sharks are killed by us than vice versa. One rather distasteful human practice is the cutting off of shark's fins before throwing them back into the water. On the afternoon of 30 August 1991 one shark got a small piece of revenge.

A Japanese fisherman from Miyagi on board the longline trawler *Fukuya* No 38 was cutting off fins from a number of shark carcasses on the deck of a longline trawler 100 nautical miles north-east of Brisbane. Suddenly a 3 metre shark weighing 270

THE DAILY

Telegraph Mirror

SYDNEY, Friday, October 2, 1992 WEATHER: Fine, 20 degrees Phone: 288 3000 60 cents*

Noeline Donaher tells

'They've made us look like fools'

HOW SYLVANIA WATERS WRECKED A FAMILY — see pages 40,41

$60,000 Bingo TRIPLE PAY TODAY

Numbers: P74

Sir Eric Neal

Five bank chiefs quit board

By ADELE FERGUSON and ANDREW STEVENSON

WESTPAC chairman Sir Eric Neal and four fellow directors resigned in a boardroom sacrifice yesterday as the bank fought to end its worst year on record.

The five – headed by Sir Eric and his deputy, Sir Neil Currie – announced their decisions in a message to the stock market just before midday, packed their briefcases and promptly left Westpac's Martin Place headquarters.

They said they had taken their full rights and entitlements and were confident the bank had a sound basis from which it could again become a pre-eminent

Continued Page 2

Terry McCrann: P11

SHARK KILLS LONE SURFER

The spot where the lone surfer was taken by a shark yesterday. INSET: Victim Michael Docherty

By DENNIS WATT

A SHARK killed a surfboard rider as his horrified mates watched the savage 20-minute attack from the beach yesterday.

The 4.2m white pointer, in a school of sharks, rammed the lone surfer's board and dragged him under, 30m off a point on Moreton Island, north of Brisbane.

Fifteen surfers on the beach watched helplessly as the shark mauled 28-year-old Michael Docherty, still connected to his board with a leg rope, and dragged him back and forth over 50m.

They saw Mr Docherty and his board emerge once but go under again as the shark thrashed about off North Point Beach.

Surfer Brett Provost, 25, likened the shark to a fisherman "playing with a line".

"Everyone felt really helpless. It could have been any one of us out there. We would have been out there if we had not gone away to get food."

The shark did not release its grip until Redcliffe police officer Sergeant Phil Sharpe and a holidaying schoolteacher harassed it from a

Continued Page 2

METRO TV: P103 Business: P37 Comics: P68 Crosswords: P58 Lottery 4777: P90 TeleClassifieds P69 (288 2000)

The Telegraph Mirror *headlines on* Friday 2 *October* 1992

kilograms latched onto him, almost severing his arm. The fisherman, 36, was winched to safety by an Army helicopter and underwent several hours of microsurgery at the Royal Brisbane

Hospital. He told doctors that he was embarrassed by the drama that his mishap had caused.

Sometimes people disappear in the water and it is never known what happened to them. When police gave up the search for Robert Bullen around Dingo Reef off Townsville in northern Queensland, they said they believed he had been taken by a shark. A number of 2 metre sharks had been seen around the fishing boat from which he had been working on Sunday 8 April 1990.

Bullen, who had moved with his family from Townsville, only a week previously, was diving for trochus shell. Steve Woolhead, the skipper of the *Bonny Kay*, was the last person to see him. Bullen was standing in waist deep water adjusting his goggles and snorkel as though preparing to dive down again. Shortly afterwards Woolhead heard a short cry.

Although Bullen was presumed by police, who spent three days searching for him, to be a shark victim, others were not so sure. Brian Lassig, a research officer for the Great Barrier Reef Marine Port Authority had this to say. 'To have a shark swallow you whole it would have to be about 6 metres long and you don't get too many of them going onto reefs. Anything smaller you would expect there to be goggles or blood or some sort of sign.'

Michael Docherty, 28, was the last person known to have died as the result of a shark attack in Queensland waters. An experienced surfer, he was pulled under water by a 4.2 metre shark at North Point on Moreton Island on Friday 2 October 1992. Two friends who were surfing with Michael looked on helplessly as the shark savaged him for a period of about 20 minutes. He probably drowned during the attack. A white pointer, which was believed to have been in the area for about three weeks, may have been responsible for Michael's death.

TORRES STRAIT

The Torres Strait is a body of water between New Guinea and Cape York Peninsula on the southern tip of Queensland. Shark attacks are enshrined in the legends of the islanders. The populations of small islands are said to have been so intimidated by packs of voracious sharks that they have had to flee to larger island groups. Since 1900 there have been six recorded cases of fatal attacks by sharks in the Torres Strait, three on the Warrior Reefs, one at Boydang and two near Thursday Island.

Among the small islands divers gather pearl and trochus. In 1914 an Aboriginal diver known as 'Teapot' or 'Treacle' had a narrow escape from a shark near Thursday Island. His head was taken completely inside the shark's mouth. As it opened his jaws to gulp him down he withdrew his head and escaped. He remained in the Thursday Island Hospital for a month afterwards. A photograph of Treacle shows a necklace of scars.

Records of attacks in the Strait have not been meticulously kept. A series of attacks, mostly on Aboriginal, Papuan and Thursday Island divers, was recorded in the 1930s by T.C. Roughley. There was a fatal attack in 1934 and another in 1935.

In 1918 Iona Asai, a diver from Saibai, just south of New Guinea, had been attacked and injured at a reef near Cairns. There are extremely few people on record to have been attacked more than once by a shark. However one Friday in 1937, at around 11 am Asai was diving in 3.5 metres of water from the lugger *San*, operated by the Protector of Aborigines, near Mabuiag (Jervis Island).

'The third time I dive and walked in the bottom. I went behind a little high place. The shark was on the other side. I never saw him and he never saw me. I saw a stone like a pearl shell on the north side and when I turned I saw the shark 6 feet away from me. He opened his mouth. Already I have no chance of escape from him. Then he came and bite me on the head. He felt it was too strong so he swallow my head and put his teeth round my neck. Then he bite me. When I felt his teeth go into my flesh I put my hands round his head and squeeze his eyes until he let me go and I make for the boat. The captain pulled me into the boat and I fainted. They got some medicines from the Jervis Island schoolteacher.'

The massive wounds around Asai's neck and shoulders required almost 200 stitches. His neck bore two rows of teeth marks. Three weeks after he left hospital he developed a small abscess on the back of his neck which discharged the broken tooth of a tiger shark.

CHAPTER TEN

▲▲▲▲▲▲▲▲▲▲▲▲▲

Victoria

In the early 1950s a young German migrant woman spent an entire night swimming in Port Phillip Bay. The fact she was alive and well — at least physically — next morning, underlined the comparative safety of swimming off Victoria's shores, the state with the least shark fatalities. In spite of the fact the number of attacks in Victoria can be counted on the fingers of one hand, people opening the Melbourne papers the following morning shook their heads and shuddered at the thought.

Some of the most enormous sharks ever caught off this continent were hooked in Victorian waters. Two giant specimens of the great white shark were caught near Brighton last century, one in 1873, the other four years later. The latter contained the complete carcass of a large Newfoundland dog in its stomach. In an address before the Field Naturalist Club of Victoria in 1889, naturalist A. H. Lucas described how some years before, strolling down Swanston Street he was attracted by a placard announcing a 36 foot (11 metre) shark on view within. This may be the shark, caught off the Victorian coast, whose jaws were on display in the British Museum for so long. It is now believed that the size of this shark was greatly overestimated. A 7 metre white pointer taken off Warrnambool caused quite a sensation in March 1909.

A tombstone in Melbourne Central Cemetery recorded that the young man buried beneath had been killed by a shark while bathing in February 1876. He apparently died on the beach at Albert Park. No one knows precisely where he was when he was mauled but there is no record of a shark ever attacking a bather body-deep in the crystal waters of Port Phillip.

The complacency of Victorians however, was shattered at 4.30 on Saturday afternoon 15 February 1930. More than 100 people had gathered at Middle Brighton to see the interstate dinghy race held by the Royal Brighton Yacht Club. They had a grandstand view of the tragedy.

Norm Clarke, 19, of North Brighton dived off the end of the pier. He dived again. This time he paused 3 metres from the platform, contentedly treading water. Someone shouted, 'Shark!' He obviously heard; he grinned and half raised his hand to wave when there was a sudden flurry in the water. He uttered a single word, more like a gasp, 'Ohhhhh!' and disappeared. A woman screamed. There was a gasp of horror when an enormous grey hulk loomed out of the water with the young man struggling frantically in its jaws like a poster for a monster movie. In an instant both shark and victim disappeared. No one could say how big the shark was or what species. Naturalist David Stead believed it must have been a great white shark, the white death, especially because the body was carried away.

On 1 December 1936, at three in the afternoon, an oar was washed up on the beach at Parkdale, about 24 kilometres southeast of Melbourne, on Port Phillip Bay. It was an omen. The oar had drifted 6.4 kilometres. It landed in the very spot from where the boat had set out the previous day. Charles Swan, 50, a returned soldier, of Parkdale, had gone off fishing in a 2.4 metre dinghy on Monday. Late in the afternoon of the following day an RAAF seaplane from Point Cooke spotted an empty, partly submerged dinghy with a 4.8 metre shark cruising nearby. The dinghy was towed back to Parkdale. There was a large hole torn in the side and shark's teeth embedded on the planking. It was ascertained that the oar must have taken three hours to drift the distance. Whatever happened out there in the Bay took place around noon. That is all that ever will be known.

The shark attack which stands out in the memory of many Victorians took place off Portsea near the entrance to Port Phillip Bay on 4 March 1956. There was a carnival abroad on Portsea Beach on that Sunday. Crowds from all over Victoria had flocked in to see the Surf Lifesaving competitions. By 4.45 pm the heats were over. The crowd had thinned. A handful of swimmers reluctant to leave the refreshing waters of the Bay lingered on. One of them was John Patrick Wishart, 26, a powerful swimmer and a member of the Surf Lifesaving Association. John's wife and sister were still on the beach when he swam to several of his mates 230 metres out.

Sharks were the last thing the six young men were thinking about, strung out, roughly abreast, treading water as they faced out to sea waiting for the next wave. One of the men was Jack Hopper, captain of the Portsea Club. He suddenly caught sight of an 'enormous black shape coming from behind us.' It skidded around in front of him (the wash was like a kick in the stomach) and dived diagonally at John Wishart. It crashed into the young swimmer and without a sound he disappeared from sight. Another of the swimmers, Greg Warland, actually saw the shark's jaws lift out of the water. There was 'one great splash' as it seemed to come down on top of Wishart. There was nothing the men could do except swim furiously for the shore.

The debate was on as to why John Wishart was taken in preference to the others. After all, the shark had bypassed Jack Hopper. One theory put forward was that the other five men were heavily tanned. John Wishart's body was pale in comparison.

Deeds and not words, however, were the order of the day. The shark hunt was on. In the following dawn a small armada put out from Portsea and Queenscliff. Fishermen anchored 4-gallon drums in a wide arc around the death scene, each with huge hooks baited with bullock's liver. Lookouts scanned the Bay from boats and the shore. This united community effort was not in vain. A shark returned to circle the area. It was spotted five times next day. On the Tuesday, 70 gallons of bullocks' blood were poured into the sea near the baited hooks. Nothing happened. A big surf rolling on the Wednesday kept the boats away but a continuous vigil was maintained from the bluff at Portsea. Then something happened. At 11 am an excited spotter reported one of the drums bobbing about furiously.

A 5.5 metre skiff with a crew of five took off from Sorrento. Several hundred people gathered to see it plough through the thunderous seas on a 14 kilometre run to the line. As the boat approached the shark's struggles became more violent. It took a further 20 minutes before it was eventually killed with a 12-bore shotgun and five bullets from a .22. The radio news bulletins brought well over a thousand people out on that Wednesday to see for themselves that the killer was really caught. No trace of John Wishart was found inside the shark. In fact its stomach was empty. It was believed it had disgorged the contents during its struggles on the hook. Whether that was the story or not, the people of Victoria were determined to believe they had caught the right shark.

A heavy shower of rain brought disaster to a young Victorian lifesaver. Chris Holland, 19, of Fitzroy, was staying at his parents' holiday cottage at Fairhaven, near Lorne, one of Melbourne's loveliest resorts, 130 kilometres from the city. On Sunday 29 January 1959 he was patrolling Fairhaven Beach with a group of young lifesavers. About 2.30 pm it began to rain, heavily. The crowd went scurrying from the sand. With the beach deserted, the lifesavers decided to 'crack a few waves' before calling it a day. His mates were several metres away when the shark attacked Chris Holland. It was brief and savage. Chris's clubmate, Russ Hughes, 18, of Preston, brought the unconscious boy in on his surfboard and he was rushed 50 kilometres in a critical condition to Geelong Hospital.

An incident which happened in reverse occurred at Melbourne's favourite, Black Rock Beach, on the day after Boxing Day 1959. A group of young children were taking swimming lessons when one of the spectators on shore spotted some spectators out at sea: three sharks. One was moving towards the children. The water was cleared in seconds and the sharks cruised the area, on and off, for six hours.

Fortune was on the side of Joe Englisch of Lower Ferntree Gully, in 1959. It was the day after naval rating, Brian Derry, was taken by a shark pack off Safety Cove, Tasmania. Englisch was swimming at Safety Beach near Dromana, on Victoria's Mornington Peninsula. A group of children on the beach spotted a dorsal moving steadily towards him. He hadn't seen it. The children screamed and shouted to him. In his dash to the shore he had to scramble over a sandbank. The shark, 3.5 metres, came after him. It got within 4.5 metres of the beach before it grounded itself in the shallows. It thrashed about before it managed to free itself.

Uninhabited Lady Julia Percy Island on the south-east corner of the continent is a favourite spot with Victorians who come to see the seals and purchase fresh crayfish from local fishermen. The waters off the island were chosen by a party from the Victorian Aqualung Club for a day's diving on 26 November 1964. Seals were abounding that day. They barked and dived along with the laughing club members. Abruptly, and for no apparent reason, the seals seemed to vanish into thin air. Suddenly something slammed into

Call of the sea was strong for marine photographer, Henri Bource *who was soon back in the water after a shark took his leg in* November 1964.

one of the divers, 25-year-old Henri Bource. Gasping for air he felt himself dragged to the floor of the ocean by a dark, bulky shape, the speed tearing off his mask and snorkel. He was in great pain and he was drowning. The shark had snapped its jaws over his leg and was shaking him like a doll. Then the shark let go. Bource came to the surface. He gulped the fresh air. He was free. But at a price. The shark had taken off his left leg below the knee.

Almost at once, Bource's girlfriend, Jill Ratcliffe, plunged into the sea from the deck of the boat, clutching a piece of rope. While Fred Arndt and Dietmar Kruppa held the stricken man, beating the water with their metal handspears, Jill Ratcliffe passed the rope around Bource's body. On deck, they held off some of the bleeding with a tourniquet as the boat ploughed back to Port Fairy while Bource's blood group was radioed ahead. An ambulance took the unconscious man to Warrnambool Base Hospital and immediately his condition was posted as serious. He had lost about 3.5 litres of blood.

Fortunately Henri Bource survived and was back in the water in weeks. In 1968 he assisted in the making of a shark documentary *Savage Shadow* in which the entire episode was simulated. He holds no hatred for sharks. 'They do what nature intended them to do and they don't kill as indiscriminately as man for the sheer pleasure of it.' Later Henri Bource specialised in diving for off-shore oil companies.

Mark Jepson, the last victim of a recorded shark attack in Victoria, was one of the lucky ones. On 10 March 1992 17-year-old Mark was surfing 200 metres off the back beach at Port Lonsdale, west of Melbourne, when he noticed a dark shadow under his board, which he at first thought was seaweed. 'I just felt it come up and munch my leg and snap my board in half,' he said. 'I didn't really want to look around to see. I thought my leg may not be there any more. I was panicking heaps, actually — pretty terrified. I never thought anything like that would happen to me.' Mark fought the shark off with his hand. To his relief it backed off and didn't come back for another try. The alarm was raised and Dr Geoff Allen was called to the beach. He found the boy lying in his torn wetsuit on the beach. He was quite calm, despite the deep wound to his right leg and hip and a badly lacerated hand. He was placed on a malibu board in the back of a utility and driven to the nearby Queenscliff Community Health Centre before being taken to Geelong Hospital for surgery.

CHAPTER ELEVEN

Tasmania

The first shark attack in Tasmania in 100 years occurred, ironically, at Safety Cove, near Port Arthur, about 94 kilometres south-east of Hobart, on Saturday 17 January 1959. Brian Derry was one of a group of young naval ratings who came ashore from HMAS *Cootamundra* for swimming and recreation. While the sailors gathered on the beach for the launch to take them back to the ship, Brian Derry decided to swim the distance. Everyone said it would be quite a feat. They never considered sharks.

T.G. Briggs, a farmer, saw what happened 270 metres off the beach. 'The water swirled in a sort of explosion. We heard the sound from the shore. Several shark fins criss-crossed the area where the rating disappeared.' Those on shore waved their clothes frantically to attract the ship's attention. After a while the launch came scudding out towards the beach. Dozens of rifle rounds were fired at the sharks, some of whom were longer than the 4.2 metre liberty boat.

In January 1974 a shark chased the three young McGuiness children on to the beach at Bruny Island, a 35 kilometre ferry run from Hobart. The determined shark chased the screaming children right on to the sand. Its head, in fact, was on dry sand, writhing and snapping, when their father and another man clubbed it to death with a paddle.

On Saturday 28 February 1982 a party of four women and five men, all members of the Reform Church in Hobart, walked along a bush track from Eagle Hawk Neck to an isolated beach at South Cape Bay on Tasmania's South-East Cape. Between them they had one snorkel and one wetsuit which they intended to take turns in using. Geert Talen, at 32 the oldest member of the group, swam out quite a distance. The others became concerned and beckoned him back. He turned and began swimming towards the shore. When he was about 25 metres from the beach Jody-Anne Wanders, 17, who was standing on the beach, saw a shark fin near Talen. The next thing she knew there was a splash and blood appeared in the water. Talen had disappeared. A few hundred kilometres away a crewman on the fishing boat *Pisces* 5 had seen a 5 metre white shark just minutes before Talen was attacked. The crew helped members of the group to search for Talen. His body was not found.

Jamie Mison, a 40-year-old abalone diver with over 20 years of experience had had many encounters with sharks, but on 14 October 1984 he underwent the most terrifying experience of all when a shark bailed him up as he hid in a rock hole for an hour. Mison left his launch near the Lanterns, off the Tasman Peninsula on Tasmania's east coast and swam down to a large clump of abalone about 50 metres away. Then he saw a huge shark beside him. 'I saw the head and teeth beside me first. It was a great white. You can't mistake them. I hit the deck, found the nearest hole and dropped into it.'

Fortunately Mison's air supply was coming by hose from his launch. On board, his deckhand, Murray Thomas, could see him through a glass bottom viewer. Mison waited for an hour while the shark stayed close by. 'I was wedged into one hole and it came in over the top of the rock. I could have reached out and touched it,' he said. Finally the shark seemed to have given up waiting. Slowly Mison edged toward his boat. To test the shark he sent up his bag which contained about 25 kilos of abalone. Murray Thomas gave him the OK signal and he swam to the boat where Thomas pulled him aboard. It was not a moment too soon. The shark reappeared right beside the boat.

After his attack Mison said he wanted to publicise his experience so that amateur divers could be made aware of the danger in the waters in the area. However it was to be another nine years before Therese Cartwright became the next fatality in Tasmania.

At 1.30 on the morning of Monday 10 February 1992 19-year-old Wayne Fitzpatrick was found in waist deep water at Clifton Beach south-west of Hobart by relatives. Wayne had been missing for six hours and was suffering from hypothermia. While paddling 150 metres from shore on Sunday night a shark had taken his malibu surfboard. His wetsuit had 15 slashes 45 centimetres long on it.

CHAPTER TWELVE

▲▲▲▲▲▲▲▲▲▲▲▲▲▲▲▲▲

South Australia

The first attack recorded in South Australia was in 1836, the year the first settlers came ashore at Holdfast Bay. Port Pirie, in the Spencer Gulf, 230 kilometres north of Adelaide was the scene of the next recorded attack. In 1884 a young girl was savaged to death by two sharks. Forty-two years passed and then an attack accurred once in each decade: 1926, 1936 and 1946. There was a 15 year gap and then a spearfisherman was attacked in remarkably similar circumstances each year for three years running from 1961.

Kitty Macully was well-known in South Australian swimming circles until she went off and married the manager of Wirraminna Station, north of Port Augusta, and became Mrs Whyte. In March 1926 she was back at the beach resort of Brighton, near Adelaide, on holiday. While her husband remained back at the station, she put her time to good use by giving children free swimming lessons. The Harbour Board had granted Kitty Whyte a special permit to swim from the end of the jetty. At about 3.45 on the afternoon of the 17th a group of youngsters had gathered at the end of the jetty. They were watching Mrs Whyte 25 metres on the Glenelg side from the platform.

The calm waters around her suddenly whipped into a turmoil. She screamed and the horrified spectators saw the body of a black shark thrashing around her. Two men came hot-footing it along the jetty and jumped onto one of the boats. As one untied the rope the other searched frantically for the rowlocks in which to place the oars, but someone had stolen them. There was a paddle in the bottom of the boat and with one paddling with his hands, they made their way to where Mrs Whyte had disappeared from view. Then she reappeared floating on top of the crimson water. The shark seemed to dive away as they approached.

The two men managed to lift her torn and unconscious body on to the boat and paddled frantically for the beach where she was promptly wrapped in a blanket. It was not realised this extra warmth was of no help to the patient. Increased body temperature is bad for circulation in a bleeding victim and there is the additional danger of infection.

Not that it mattered any longer. By the time she reached the Bindarr Private Hospital on the Jetty Road, Mrs Whyte was dead. Soon after, the shark reappeared and circled beneath the bloodstained boat. It hung about for some time and repeated attempts to hit it with rifle fire failed. Adelaide was badly shaken. They had learnt for the first time that their favourite beaches were no longer completely safe.

Most people had gone home for their tea by 6 pm, but 18-year-old Len Bedford lingered on West Beach, near Adelaide — midway between Henley Beach and Glenelg — that Wednesday, 22 January 1936. A scream sounded from the water. He looked out to a spot 270 metres from the shore. 'I saw a bather struggling. Then I saw a large shark flash out of the water, its tail sticking up in the air, and the bather disappeared.' He ran to the water's edge. There was nothing in sight. Small waves splashed around his feet. Turning, he spotted a solitary bicycle lying on the sand. On top of it was some boy's clothing and leather bag.

At that moment a sulky came trotting past. Len called out to the man and told him what had happened and the man promised to inform the police while he waited on the beach. But the man never did bother to call at the police station. After waiting for more than an hour, Len Bedford borrowed the bicycle — it obviously belonged to the victim for there was no one else in sight — and peddled to Glenelg as fast as he could. Inside the coat was a notebook. It read: R. Bennett, 28 Lombard Street, North Adelaide. There were lists of names and amounts written beside them. Raymond Bennett was a 13-year-old newsboy who had gone to the beach after he made his paper sales, to cool off in the sea. He was never seen again.

On Thursday 21 January 1960 Mr S. Smith of Rendlesham, SA, was in a dinghy with two 14-year-old schoolgirls, off the coast near Millicent in the south-east of South Australia. The girls were laughing at a seal, barking and showing off near their 4 metre craft. The seal's games came to an abrupt end when a shark, 'bigger than the boat,' attacked and killed it.

Mr Smith promptly started up the motor and began moving

away from the scene, but the shark came after them. The two girls huddled together, terrified, as it bumped and brushed against the boat, throwing the occupants all over the place. They continued to make for the shore but the shark would not leave them alone. It swam beneath the boat and the two girls screamed as they were lifted almost clean out of the water. Their ordeal lasted almost half an hour with the shark swirling and side-swiping them all the way to the shallows.

An incredible incident underlining man's impotence in the shark's domain eventuated during the South Australian spearfishing and skindiving championships off Aldinga Beach in St Vincent's Gulf, 90 kilometres south of Adelaide in December 1961.

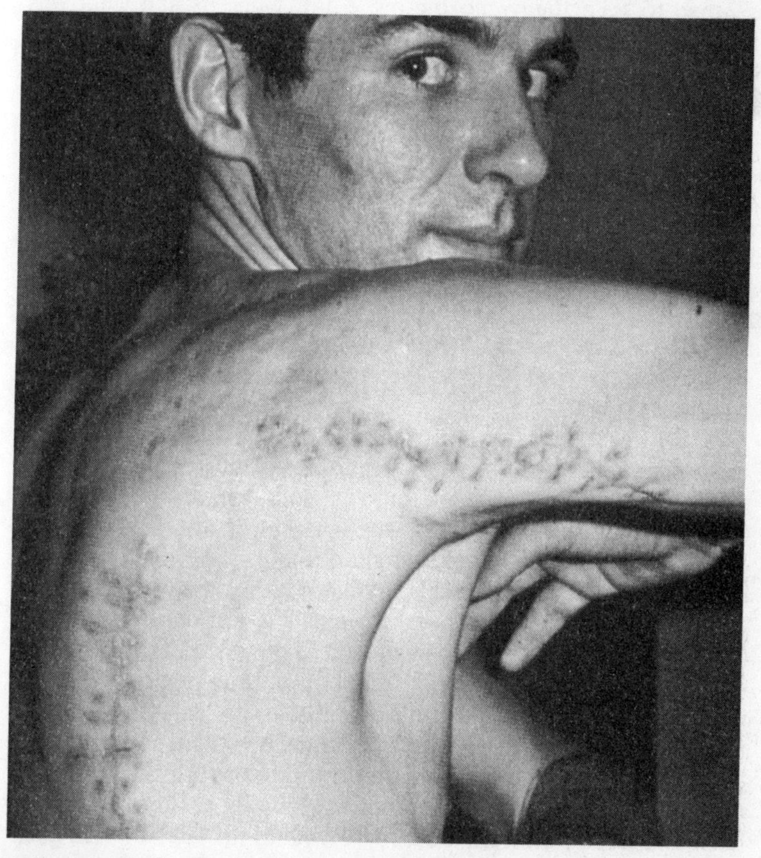

Rodney Fox after being in the jaws of a shark. 'The ultimate horror was it was my master.'

Rodney Fox, 27, dived after a 9 kilogram morwong. He gives an illuminating word-picture of the undersea world. 'How can I describe the sudden silence? It was perceptible hush even in that quiet world, a motionlessness that was somehow communicable, deep below the surface of the sea. Then something huge hit me with tremendous force on my left side and heaved me through the water. I was dumbfounded. Now the "thing" was pushing me through the water with wild speed and I felt a bewildering sense of nausea. The pressure on my back and chest was immense. A queer, "cushioning" feeling ran down my right side as if my insides on my left were being squeezed over to my right side. I had lost my face mask and I could not see in the blur. My spear gun was knocked violently out of my hand. The pressure of my body seemed to be actually choking me. I didn't understand what was happening. I tried to shake myself loose but found my body was clamped as if in a vice. With awful revulsion my mind came into focus and I realised: the shark had me in its jaws.'

Suddenly those jaws relaxed. For the moment he was free. He jerked backwards and pushed himself away. His right arm went straight into the shark's mouth. He felt pain and fright. Now he struggled desperately to the surface, bumping repeatedly against the shark's brittle body. The welcoming blue sky hit his eyes and he gulped air. But the big fish came after him. An icy shudder went through his bruised body as its fin grazed his flippers. In a frantic bid to avoid its jaws he wrapped his feet around its body. The shark dived so deep, Rodney Fox scraped the rocks on the ocean bottom.

With his last ounce of strength he floundered to the surface. 'The shark hit surface a few feet away ... its hideous body was like a great rolling tree-trunk, but rust-coloured with a great pectoral fin. The ultimate horror was that it was my master. I was alone in its domain. Here the shark made the rules. I was no longer an Adelaide insurance salesman. I was something to eat.'

Once again a shark was to act in a strange and unpredictable manner. It made towards him, then veered aside and disappeared. He shouted, 'Shark! Shark!' All he remembered was a voice saying, 'Hang on mate. It's over. Hang on.' He believes those rough, comforting words saved him from dying there and then.

His right arm was laid bare to the bone, in several places. His chest, back and left shoulder were badly gashed. Pieces of flesh had been torn from his body. Five months later he was back in the water. He gave up selling insurance and became a professional abalone diver, right in shark territory, off the South Australian coast.

Three years and nine months earlier, again at Aldinga, on 12 March 1961 another skindiver, Brian Rodger was attacked under almost identical circumstances. Only he was in a position to defend himself.

The annual spearfishing championships were well under way. It was close to lunchtime and although he had been in the water for some time, Brian decided to add to his day's tally. He was some distance from the beach when his intentions were dramatically interrupted. In a moment his anguished body was in the tearing grip of a dreaded white pointer. He lashed out with his fist and his arm went into the shark's throat, ripping the flesh on its upper teeth. 'Surprisingly it did let go, and then came back in a fast, tight circle for another bite. That was really terrifying because now I could see the whole shark and the size and enormous power of him.' Brian was still grasping his spear gun. He let fly with a spear and to his immense satisfaction it embedded itself in the top of the shark's head. It halted its charge. It shook its head violently from side to side, then dived away into the gloom.

Brian Rodger was alone in a cloud of bloodied water. He made a tourniquet from the rubber of his spear gun. His arm and his leg were severely lacerated and he was a long way out. Now began the slow agonising swim to the faraway shore. His strokes became weaker and weaker through loss of blood. The shoreline seemed an interminable distance. 'The land, beautiful land, came closer. I waved and yelled, "Shark!" ' There appeared to be no reaction from the figures moving about the beach. Hazily he wondered whether he was going to make it. Then, from nowhere, a small rowboat appeared with two young men.

There was no chance of the brawny skindiver fitting in the boat with the pair of them. 'Hang on!' One of the men jumped in the water, regardless of whether the shark was still lurking and helping to heave Brian into the boat, he swam behind, to push the boat faster. A score of divers came running across the reef to meet them. The boat was lifted bodily and deposited gently on the beach. Police cleared the way for an ambulance-dash to the Royal Adelaide Hospital. Skilled surgery, including 200 stitches, and Brian's excellent physical condition, got him through. A year later he came second to Ron Taylor in the Australian Spearfishing Championships over in the west.

A year and nine months later the story was again repeated. This time there was no happy ending. On Sunday 10 December 1962 a spearfishing competition was in full swing 20 kilometres south of Aldinga, at Caracalinga Head. Two young men were some way from the shore, about 180 metres and in deep water. They had a surf ski between them.

Allen Phillips had just surfaced from a dive and saw a splashing near his friend Jeff Corner. The tail of a shark broke water. It appeared to have Jeff in its mouth. Managing to overcome his sense of dread Allen scrambled on to the ski and paddled toward his stricken mate. He reached out for the unconscious 16-year-old but the shark — again a white pointer — jerked its prize from his grasp. The shark, still clutching the boy, surfaced the other side of the ski.

Allen Phillips banged it with the paddle as hard as he could. Surprisingly it let go. Jeff Corner's eyes were mute and expressionless as his friend raised him up on to the ski. His leg was a terrible sight. Allen Phillips threw his catch to distract the white pointer which edged slowly a short distance away as though it were watching the scene. It ignored the fish. Jeff Corner's limp body was half on the ski, being held up by his friend. The shark moved slowly with them, its eye fixed on them. Any moment it might decide to charge.

Another spearfisherman came paddling furiously across to them. His name was Murray Brampton and he yelled at them to keep going. Out of the corner of his eye, Allen Phillips saw Brampton bashing at the shark with his paddle. When Murray Brampton joined the crowd on the beach, including the boy's parents, Jeff Corner was already dead.

Two savage attacks along the same strip of South Australian coast which occurred in the 1970s emphasise the vulnerability of even the most experienced divers. On 9 January 1974 abalone diver, Terry Manuel, 26, a married man with one child was attacked near Streaky Bay, 730 kilometres north-east of Adelaide. He had been down for half an hour. Suddenly he surfaced screaming, 'Shark!' He was surrounded by blood. The shark, probably a white pointer, continued to maul him even as it dragged him under. He was dead by the time his mate, John Talbot, got to the spot in their dinghy.

Thirteen months later, on 10 February 1975, at Penong East off Streaky Bay, Wade Shippard was swimming underwater from a crayfish cage to a fishing bait when a 2.4 metre shark took his right leg. The incidents were reminders of the risks taken by divers such as the Ben Cropps and Valerie and Ron Taylors, who rub shoulders with sharks for a living.

On Monday 29 December 1975 the shark patrol plane reported a shark off Glenelg, Adelaide's favourite beach. The rescue boat of the Glenelg Surf Club put out to chase it away. It charged the boat, slamming against the bow. It was a 2.4 metre shark, big enough to be a serious menace to humans, but not big enough to do much damage to a heavy surf boat. Nevertheless just when the lifesavers thought they had forced it out to sea, it doubled back and banged them again. 'It charged us a couple more times,' said the boat's captain, G. W. Scarfe, 'the last time was in water so shallow the jet boat was bottoming quite a lot'. When a shark is aroused nothing will put it off.

In September 1976 a professional abalone diver warned that the South Australian Government should be investigating the possibility of meshing Adelaide beaches if they were not to become 'one long smorgasbord' for increasing numbers of white pointers in South Australia's gulf waters. High mercury levels found in sharks had stopped fishermen going after them, Mr J. R. McGovern said, and in the past 18 months he had seen more sharks than ever before in six years of diving and many years working on crayboats.

McGovern's warning came just days after a 27-year-old skindiver was attacked at Point Lowly, north of Whyalla. Darryl Richardson was diving with his brother on Sunday 19 September near rocks about 40 kilometres north-east of Whyalla, when a 2.7 metre shark bit both his legs in one bite as it swam across his body. Bleeding profusely, he struggled onto nearby rocks. His thick, tight-fitting wetsuit together with the prompt action of his brother in applying a tourniquet probably saved him from bleeding to death. There were deep puncture wounds in his right leg below the knee and around the ankle and foot of his left leg. Darryl's brother alerted a nearby lighthouse keeper and began to drive his brother to hospital. Meanwhile the lighthouse keeper called an ambulance which met them on the road. Darryl was taken to Whyalla Hospital. As the ambulance officers said, he was lucky to be alive.

Almost exactly one year later, near Ceduna, in the Great Australian Bight near the West Australian border, a young surfer holidaying from Perth was attacked. Philip Horley, 17, entered the water at Cactus Beach at about 4 o'clock on a Wednesday afternoon late in August 1977 with about 12 other surfers. He was keen as there had been no waves for the past four days.

Philip was paddling with a mate about 300 metres off shore when a large white pointer rammed his board, knocking him into the water. The shark brushed across his body. 'I was looking the shark straight in the eye-ball and I thought I'd had my day.' He turned and began swimming towards the shore. Fortunately the shark seemed more interested in his board than in Philip. 'It started munching it as if it was in a frenzy.'

Because the water was so cold Philip did not immediately realise that he was injured. As he swam to shore he felt his leg and found that it had been gashed wide open. There were four 12 centimetre gashes across his left thigh. The cold also helped to stop the bleeding. On the beach about 30 surfers gathered around to apply a tourniquet to Philip's leg and head for the hospital in Ceduna.

For the witnesses of some attacks the horror may never really fade. It is especially traumatic when the victim is a close relative. The family of Shirley Durdin will never forget what they saw at 12.30 on the afternoon of 3 March 1985.

Shirley, 33, was snorkelling for clams with her husband, Barry, and a friend in 2 metres of water at Wiseman's Beach in Peake Bay, about 35 kilometres north of Port Lincoln on the southern tip of the Eyre Peninsula. Her four children, a boy and three girls aged between five and 11, were watching from the shore 150 metres away.

Another witness was Kevin Wiseman, a retired fisherman who was standing on a cliff above the beach. 'There was a big spray and red froth and bubbles — a mass of blood in the water. One chap was yelling, "Help, help, help, she's gone, she's gone." I could see the fin and tail from where I stood and it was a lot longer than the boat that went out to help.'

Kevin Hirschausen had just brought his boat in from fishing and was standing on the beach with his daughter. Seeing the splash and a shower of red water they quickly got back into the boat and headed back to help. Shirley had been bitten in half. Now the 6 metre white pointer came back to tear her head and an arm from her legless torso. As the boat approached the Hirschausens saw the shark come back. 'Oh, Christ, it was awful,' Hirschausen said. 'As we got near we saw him come up and grab the rest and dive down, chewing. The arm was sticking out from its jaw. It was gruesome. We saw her flipper, but we didn't stop out there. The shark was still cruising and it was too big for my boat.'

Barry Durdin was still in shock two days later. The effect on four impressionable children can hardly be imagined. The attack on Shirley Durdin was the first case in Australia where a victim was actually known to be eaten by a shark.

Terry Gibson was probably also bitten in half by a white pointer, but there were no witnesses to see this attack on Friday 18 September 1987. Terry, a 47-year-old father of six, had been diving alone for scallops off Marino Rocks in Adelaide. A diver with over 20 years experience, he ran a diving business at Christies Beach and was well-known among the Adelaide diving set.

A search began for Terry on Friday afternoon after his empty aluminium boat was found anchored near the Marino Rocks boat ramp. Fishermen reported seeing heavy splashing and seagull activity at about the time of Terry's disappearance. That night a face mask and three bags of scallops were found in the water. The search continued early the next day. Shortly before noon an air tank, attached to a ripped buoyancy vest and a diver's weightbelt, was found on the seabed by police divers, bearing teethmarks. The belt was in the locked position and could only have got to the bottom if it had slipped off the diver's torso. The search was called off.

Lynette Gibson, Terry's wife, appealed to people not to go out shooting sharks. Terry had loved diving and been aware of the risks. 'We're not against the sharks,' she said.

Waitpinga Beach near Encounter Bay was the scene of another fatality on the night of Wednesday 11 March 1989 when two young surfers encountered a 2 metre shark. Matthew Foale died after being savaged by the shark despite the vain efforts of his mate Trevor Viney to fight it off.

An underwater reef reserve 350 metres from Aldinga Beach is a popular diving spot, despite a number of shark attacks in the area. On Sunday 8 September 1991 a group of eight divers took advantage of warm weather and clear skies to dive on the reef. They took out a boat and descended under the water. Dave Roberts, the senior diver in the team, said he 'heard a great thunder' which he thought was a boat. He turned to see a white pointer, probably about 3 to 4 metres long. It thrashed past quite close to him. The water was quite murky and other members of the team didn't see the shark until it was right on top of a 19-year-old Adelaide student named Lee. They could do nothing but scramble into the boat and light a flare to raise the alarm. All that was found of Lee after a search by police in boats and a rescue helicopter was his fins and an undamaged tank.

Lee's fins and undamaged tank are displayed by police officers.

CHAPTER THIRTEEN

Western Australia

For many years it was believed that the cold current which sweeps along the coast of Western Australia from the West Wind drift of the Antarctic provided partial protection from shark attack. The record of shark attack in WA is relatively sparse compared to that of Queensland and Victoria. Then again, the population is much smaller.

There have been 24 recorded attacks on swimmers in WA. Of these eight have died. WA's first recorded shark attack was in 1803. Three people were mauled in the north of the state, two of them fatally, in the 19th century. There were probably many more attacks on pearl divers which were not recorded.

In January 1923 13-year-old Scotch College student Charles Robertson was taken while swimming in the Swan River 25 metres from the Claremont shore. Later that year, in November, two Koepang Islanders were killed while dry-shelling off Condon.

Cottesloe Beach is sometimes known as Perth's Bondi. Perhaps it is inevitable that Cottesloe should be the scene of Western Australia's classic surf attack. Its sidelight provided a terrifying example of the killer shark's single mindedness.

Imagine Cottesloe in the mid-twenties, a golden Sunday afternoon early in the season. The beach was crowded and so was the water. The word 'shark' is frequently used in fun but when it sounded over the waves and the din of voices soon after three o'clock, it had an ominous note. There was a mass dash and splash to the shore. A group of lifesavers were pushing out a dinghy when they caught the word. They looked up and a figure, now isolated from the retreating bathers, was riding face down in the water. Already three club members were breasting the waves now flecked with red. Not for one moment had these young men considered their own safety.

The shark had selected its prey from among the dozens of surfers. As the bloodied victim was being hoisted into the boat the shark barged between his rescuers and took another bite at his trailing leg. The stricken man spoke his only words, 'Oh my God.' The lifesavers banged the shark with a heavy oar. It released its hold, moved off several metres then turned about, determined not to be cheated of its prey and came full tilt for the boat, its head appearing over the side as it slammed in, almost capsizing it. Several people, watching the drama from the beach became hysterical. Dr Harpur of Fremantle Hospital was on the beach but there was little he could do.

Fate had chosen Samuel Ettleson, 55, a bookmakers' clerk, to be plucked from amid the crowds of men, women and children and to die by the jaws of a shark on that afternoon, 22 November 1925. For hours after the 3.6 metre shark kept reappearing close to the beach. Several rifle shots had little effect. That evening a well-known Cottesloe bather who had earned the name of 'Shark Bait' came trotting on to the beach for his customary dip. It was only after much persuasion by the lifesavers that he agreed not to go in. As late as ten o'clock that night groups of people still lined the foreshore and spoke in whispers.

Sunday 10 February 1946 turned out to be a wild day along Perth's City Beach. It began when an 18-year-old sailor, Ronald Sunderland, off HMAS *Leeuwin*, was attacked by a 4.2 metre shark in calm, waist-deep water off the north end of the beach at 12.15 pm. The shark took three bites, and if several men had not rushed in and carried him out, it would certainly have killed him.

Fortunately there was a doctor on the beach and while lifesavers — totally unprepared for an event that had never happened before — tore off on their motor cycles for first aid equipment the young sailor was made comfortable in a nearby tea room. In spite of the fact that because of a mix-up, the St John Ambulance did not arrive until 12.50 pm, Sunderland was still alive when he reached Hollywood Military Hospital. Immediate blood transfusion and amputation eventually saved his life.

Back at the beach a surfboat, manned by a crew from the City of Perth Surf Lifesaving Club, put out from the beach in search of the shark. A dorsal broke the water and they ploughed after it. It was only several months from the end of World War II and the charges used for their harpoon gun had gone towards the war effort. Like a scene in an old painting, one of the lads crouched in the bow, clutching a harpoon, his arm poised.

The first harpoon bounced off the shark's back. The second struck home. The shark leapt over half a metre and gave the boys a nasty shock as it rushed at their boat and bit the boards level with the watermark, leaving behind two teeth. They were in for an even nastier one. It rushed around the boat, diving and surfacing, the harpoon still in its back.

Meanwhile the lifesavers recovered the first harpoon and managed to straighten it out. By now the shark had shaken off the harpoon and as it swirled close to the boat they thrust at it again. It crashed into the craft, half its body reaching inside, snapping at the arms and legs of the crew who, fortunately, had been toppled into the bottom of the boat.

One of the men was thrown into the water and scrambled back in an instant as the infuriated shark took off in another direction. The shaken crew got themselves together, laughing nervously. But the shark was back. It torpedoed in from a short distance and smashed into the boat almost capsizing them, then it went off in the direction of Scarborough. The crew had taken enough by now and rowed vigorously back to the beach. But the shark was spotted again as its dorsal moved menacingly towards a crowd of bathers. The shark siren was sounded and the water was cleared. No one was game to enter the water for the rest of the afternoon.

The colourful days of Western Australia's pearling industry have long passed since reaching their peak in the mid-twenties. In a tough, ruthless industry, records of Asians and Aborigines attacked by sharks were never kept. The centre of the industry was Broome, 2,000 kilometres north of Perth. Today Broome attracts numbers of tourists, many of whom visit the authentic dinosaur tracks embedded in the sandstone at the base of the cliffs of Gantheaume Point. The Point commands a view of a magnificent stretch of beach. It was here on 16 May 1949 that a shark, whose ancestors roamed the seas when the dinosaur first made those tracks, attacked Mary Passaris, a typist in her early twenties. The attack had shades of Sydney's Shark Arm Case. The shark took Miss Passaris's left forearm before she was dramatically rescued by Mrs L. M. Maxwell and Mr L. Pearce, who later received awards for their bravery.

While Mary Passaris lay in hospital a determined vigil was kept for the killer by local residents. Baits were set at the end of the jetty and fishermen kept a sharp lookout. On the fifth day following the attack, their patience was rewarded. A 2.7 metre blue whaler was caught but not before it snapped the centimetre hook and had several bullets pumped into its body. The missing arm was found preserved in the shark's stomach. Miss Passaris's ring was still on a finger. It is possible the glint of the ring may have attracted the shark in the first place. Mary Passaris recovered and it is said she continued to wear the ring, on the finger of her right hand.

In most instances a shark is so single-minded in its attack on its prey, it will completely disregard all rescuers, but not always ... One such episode occurred during a spearfishing competition held at Jurien Bay, WA, 200 kilometres north of Perth on 21 August 1967. Robert Bartle, 23, had dived to pick up a float line when he was attacked.

Lee Warner, 24, a schoolteacher who held the state's spearfishing championship five times, was swimming eastward when he glimpsed the long shadow several feet below him in the opposite direction. 'It moved so fast, by the time I looked back it had Bob in its mouth and was shaking him like a leaf. I dived straight down and put a spear in the top of its head; right where the brain should have been. It didn't seem to affect it. The shark started circling and came at me. I only had an empty gun.' Warner attempted to hit the shark in the eye with the gun. 'It began circling again, keeping about 8 feet away. I realised I was helpless without a spear. I got Bob's. It was floating only a few feet away. I could see its jaw was much wider than the body. The jaw must have been 2 feet wide.' Clouds of blood obscured his vision. He aimed the spear at the brute's eye, but missed. A new threat flashed into view: a bronze whaler. As it happens it probably saved his life for it distracted the killer shark.

Lee Warner knew Bob Bartle was already dead. He swam, backwards, most of the 130 metres to the shore. He bounced into his car and raced 10 kilometres for help, returning in a boat to the scene of the disaster. The shark was still there. Bob Bartle's body was nowhere to be seen.

The upper part of Bartle's body was later recovered. He had been bitten completely in half, his legs and buttocks swallowed whole: it seemed like a classic white shark attack. Whites had previously been caught in Jurien Bay. On the other hand Lee Warner told of seeing the attacking shark's eye roll white as though covered in white skin. The nictitating membrane or third eyelid which protects the eyes of some sharks during an attack is not present in the white pointer, although they can roll their eyes back in their sockets. A tiger shark, however does have a nictitating membrane. The even teeth marks on the upper and lower surfaces of Bartle's body also suggested a tiger shark. Whatever the shark was will never be known.

On Saturday 30 August 1975 Dennis Thompson, Barry Paxman and Glynn Dromey took a boat to go spearfishing off the coast about 15 kilometres north of Lancelin, north of Perth. Thompson and Paxman were treading water and spreading berley to catch kingfish when a whaler, 2.4 metres long swam past them three times. At first they ignored it but as it came past the third time Paxman, fearing its intentions, speared it through the head. It sank to the bottom. Thompson dived after it, intending to finish it off. As he neared the bottom the shark twisted quickly and swam up towards him.

'I felt it hit my back, then I saw it, with the spear sticking through its head, tugging at my arm. At that stage I didn't care whether I lost my arm, I just didn't want to be killed. I hit out at the shark with my left hand and pulled away from it. Then it released me and I did not see it again. Though there was a lot of blood about I did not feel any pain, nor did I feel any panic till I reached the surface and realised the danger I had been in.'

On the surface the two men called out to Glynn Dromey in their power dinghy 60 metres away. Thinking they wanted him to join them fishing, Dromey jumped into the water and swam towards them. Finally they managed to get back into the dinghy and back to Lancelin where Thompson was treated by a first aid officer before being taken to Royal Perth Hospital for surgery to a severe wound between his right elbow and shoulder. Thompson, 29, despite having enjoyed spearfishing for eight years, vowed that he would give it away. 'I don't think my wife would want me to carry on,' he said.

Five years later a young sailor became the next person to be mauled by a shark at Lancelin. Able Seaman G. R. Bohan was stationed on HMAS *Stirling* at Garden Island. After a training run along the beach he dived into the water for a cooling swim. About

60 metres out something hit him as he dived under a wave. 'It stood me right up out of the wave,' Bohan said. 'That's about all I know about it. Bohan had been quite severely mauled on his right shoulder and there were puncture wounds across his chest and deep cuts and slashes on his back. Despite his wounds he was able to swim to the beach, where he was found by other sailors. An RAAF rescue helicopter flew him to Royal Perth Hospital, where he recovered.

On 16 February 1981 Steven Fawcett of Willagee was exercising his dog in shallow water at Leighton Beach, north of Fremantle. About 100 metres offshore something bit his foot and he saw a metre long hammerhead swimming away. A policeman on the beach took him to Fremantle Hospital where he was treated for deep puncture wounds.

Rottnest Island is a popular recreational spot for tourists and locals, just 30 kilometres west of Perth. Earlier in its history it was used as a gaol and housed a large population of Aboriginal prisoners. The theory was that anyone who tried to escape would be eaten by sharks and it is definitely true that there are many sharks to be seen in the waters around the island. Nevertheless, recorded attacks are far and few between and diving and fishing are popular pastimes.

The fact that sharks do lurk in these waters came to public notice in two almost identical incidents in 1986 and 1993. Ivan Anderton was fishing with a mate in a 5.4 metre boat about 3 kilometres west of Rottnest on 26 July 1986 when a huge shark passed underneath them. At first they thought it was a whale. Then it circled up and its dorsal fin broke the surface. 'It was looking at me with an eye as big as a saucer,' Anderton said. 'The next minute it came at the boat and bit the motor, lifting the stern about half a metre out of the water. We didn't even bother to lift the anchor, we just took off.' Part of a tooth was later found in the motor. Anderton, who had fished the area for years, said that the shark, a white pointer, which he estimated to be 6 metres long, was the largest he had ever seen.

Seven years later, on Sunday 15 August 1993, a similar sized shark, also thought to be a white pointer, struck another 5.4 metre fishing boat west of Rottnest, kicking up the outboard motor. 'Was it the same monster?' asked a local newspaper.

Sharks are often seen swimming close to the shore at Esperance, on Western Australia's southern coast, particularly in warm weather. Many of them are bronze whalers, not averse to biting humans, so it is surprising that there have been very few attacks in the area.

An attack on a young surfer in 1987 is thought to have been the first such attack in the town's history. Grantley Butcher was surfing about 300 metres offshore between Twilight and Fourth Beaches, west of Esperance on Monday 30 August when a 2 metre bronze whaler knocked him from his board, grabbed his leg and dragged him underwater. After a few seconds, in which man and beast wrestled in the water, Grant's ankle-strap became caught in the shark's mouth. Grant was able to paddle to shore, his leg bleeding badly. As he glanced back at the shark it was thrashing around with something in its mouth. It appeared to be in a frenzy. Grant crawled to his car and drove himself to the Esperance District Hospital, 8 kilometres away.

Mako sharks are known for their spectacular leaps from the water, making them a popular game fish. For three Bunbury fishermen in 1988 a mako's leaps became something more frightening than a game.

Steven Piggott, Kelvin Martin and Brendan Martin were drift-fishing 18 kilometres west of Bunbury, south of Perth, on Friday 12 February in a 5.5 metre fibreglass boat. A 3 metre, 200 kilogram mako, hooked by Steve Piggott, suddenly leapt into the air and onto the boat, knocking Steve and Kelvin into the water. Brendan climbed out of the way of the thrashing shark onto the bow. He was soon joined by the other two, anxious to avoid the possible presence of more sharks in the water. As the mako thrashed around on the deck, the three men radioed for help. For the next hour and a half they waited while the shark completely destroyed the inside of the boat, almost capsizing it in the process. By the time the sea-rescue squad arrived to tow them back to shore the shark had thrashed itself to death.

Chad Wittorff was not going to give up his boogie board without a struggle when a shark took a liking to it in the surf at Backbeach, Halls Head, south of Perth. Fourteen-year-old Chad was surfing about 15 metres from the shoreline in November 1991 when the shark surfaced and grabbed hold of his board in its jaws. Perhaps his decision to fight for it was influenced by the fact that the shark was a small one. However, having rescued the board from which a chunk the size of a hand had been taken, Chad was not taking any more chances. He made for the shore. The shark wasn't going to let him get away that easily. In knee deep water it made a grab for his leg, leaving long scratches on his calf. It was an unusual attack for Perth waters. Small sharks usually steer clear of humans.

CHAPTER FOURTEEN

▲▲▲▲▲▲▲▲▲▲▲▲▲▲▲▲▲▲▲▲

Shark recognition

It makes sense for skindivers and fishermen to be able to identify a dangerous shark when they see one. The wobbegong and the hammerhead are easy to pick, but they are not so dangerous. For other sharks a quick appraisal of the dorsal fin, snout, colour, markings, teeth — from a safe distance — and even mannerisms might give the answer. In surprisingly few cases in attacks on humans, has the species of the shark been named. Even length estimates cannot be relied upon. The refraction of light rays in the water tend to concertina length to the human eye. The Australian big three among the man-killers are the white shark, whaler and tiger. The hammerhead, grey nurse and thresher cannot be entirely trusted. Neither can the wobbegong. Any creature with a nickname like 'wobby' shouldn't be too dangerous, but people have died as a result of wobbegong wounds. The following are some of the sharks commonly seen in Australian waters.

GREY NURSE (*Eugomphodus taurus* or *Carcharias taurus*)
Also known as the sand shark especially in the USA, the grey nurse is Australia's most well-known and most maligned shark. Perhaps it is the name that has an appeal. 'People like the name so we give it to them', said one pressman. In fact there are very few reliable reports of unprovoked attacks by the grey nurse on humans; two exceptions are two attacks which took place in 1961 on Len McWhinney in WA and Joe Prosch in Victoria. They are more docile than other killer sharks and thus more easily caught. They appear to thrive in captivity.

The grey nurse is easily recognisable by its second dorsal

fin, almost as large as the first, its long body, usually between 3 and 3.6 metres, flattened head and sharp snout. It has long, slender, pointed teeth.

In spite of its formidable looks, the grey nurse is a slow, lazy creature. Aboriginal people in northern coastal areas frequently ignore them in the water.

Zane Grey (1875-1939), a US writer of classic westerns who did much to promote big game fishing in the south Pacific, caught six grey nurse in one day. On Sunday 17 September 1967, at the opening of the fishing season, eight grey nurse were caught with king fish off Maroubra, by the launch *Overdraft*.

Because of their tendency to hang motionless in small groups in marine gutters, grey nurse are extremely vulnerable to powerheads used by spearfishermen. As many as 30 have been killed in just one day on one NSW reef.

One year, Manly Marineland's Frank Jenkings expressed concern about the large number of grey nurse being caught. He believed their haunts might well be taken over by fiercer species. Grey nurse numbers have been reduced to the point where the species is in danger of extinction. A female produces only one or two young, born alive nine months after fertilisation, so recovery of the population is very slow. By declaring this species a protected species, it is hoped that their numbers will increase.

HAMMERHEAD (*Sphyrna ssp.*)

This nightmarish looking creature with its bizarre shaped 'hammer' head is not quite as dangerous to humans as it looks. It is a lively, nervous, swift-moving fish, often the first to appear when a tempting bait is around. There are several species, each identified by slight differences in head shape.

Hammerheads are at home in the open sea and close to shore, but have seldom been seen inside the breaker line in Australia. They like brackish water. The hammerhead prefers to swim with groups of mates but is also quick to take a bite out of them with its broad-based triangular teeth. Conversely other species seem to like the taste of the hammerhead most. This may have something to do with the fact that they have a higher potency of liver oil than any other shark.

Their eyes are at the end of the side extensions of their heads. They are brownish grey to dark olive above, a paler shade below. Their tails are huge and tower above the first dorsal fin. Some species grow up to 6 metres.

Fishermen say you never know how a hammerhead will react to the hook. They seem to die quickly. Life expectancy in captivity is less than most other species.

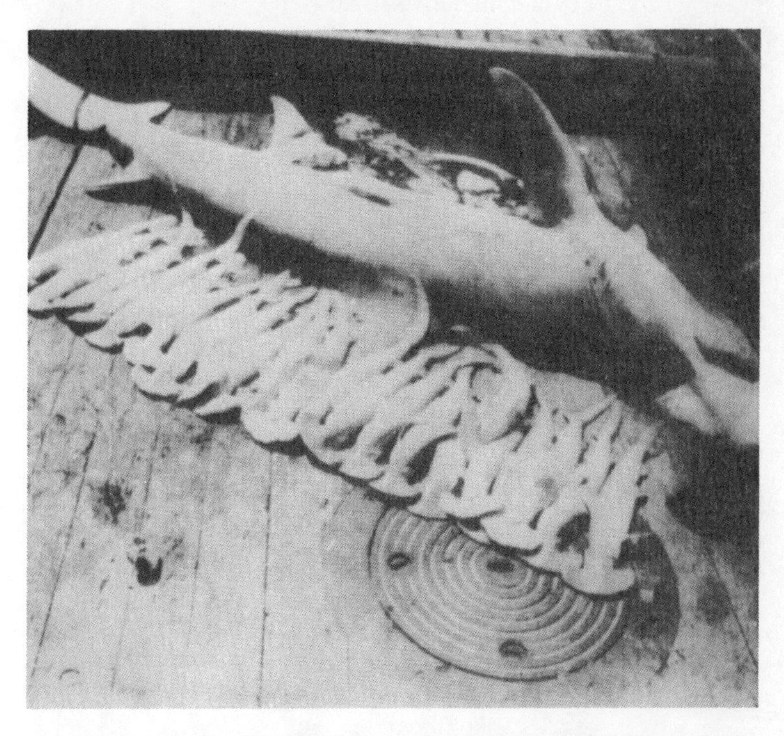

Female hammerhead and her litter of pups. Although shark litters are relatively small, the young are born ready to fend and forage. Survival rate is high.

The hammerhead's sinister reputation goes back to 1815 when, purportedly, portions of a male human were taken from the stomach of a hammerhead off Long Island. There are few reported cases of attack by this species. In fact its jagged teeth are noticeably feeble compared with the white or the tiger.

A hammerhead attacked Miss Gertrude Holaday, badly wounding her, 60 metres from the shore at West Palm Beach, Florida on Monday 21 September 1931. On a September day in 1959, Robert Walker, 29, was attacked by a hammerhead in the Gulf of Mexico while clinging to a seat cushion after his boat capsized. It pulled him under but he managed to fight it off with his hands which were badly bitten. It was quickly joined by other hammerheads but he was rescued just in time.

In the summer of 1956 Captain Cousteau was driven ashore by large schools of hammerheads in Australia.

The mako can swim at a speed of 40 knots. A beautiful and deliciously edible game fish, it fights ferociously when hooked, leaping high and sometimes rushing the boat.

SHORTFIN MAKO (*Isurus oxyrinchus*)
The shortfin mako is also known as the blue pointer or sharp-nosed mackerel shark and is closely related to the white shark.

'The mako is the aristocrat of all sharks,' wrote Zane Grey. 'His leaps are prodigious, inconceivably high. His ease and grace is indescribable. It is unfitting to call him a shark at all.'

Mako is a Maori name and the mako is often regarded as New Zealand's premier fish. It prefers the open sea where it seems to delight in attacking boats. Stories of the vicious mako biting into the wooden hull of boats, frequently leaving some of its teeth behind, have come from around the world. There appear to be few records of attacks on swimmers.

Big game fishermen have much respect for the mako. It has the reputation of a tough opponent and a gallant fighter. It has been known to shoot straight into the air, higher than a fishing launch mast, sometimes crashing down on the cockpit, wrecking everything in sight.

On more than one occasion fishermen have been scared out of their wits when the head of a mako with the 'diabolical eye of a creature that would kill as he was being killed',' appeared over the ships rail to peer briefly and obliquely around the deck.

The colour of the mako is dark, blue-grey or blue above, whitish below. Its first dorsal fin is closely followed by a very small second dorsal. It can grow to up to 4.5 metres. The mako is noted for its practice of chopping off the tails of larger fish to render them helpless.

THRESHER (*Alopias vulpinus*)

Like the hammerhead, the thresher is an odd man out in the shark species. Its tail, upon which it relies almost entirely for its tucker, is outlandish in size (in the case of baby threshers the tail wags the fish).

The thresher is more froth and bubble than bite. But its wild threshing certainly impresses the herrings or mackerel it rounds up into tight masses with its tail-whip, before moving in among them with wide open jaws. Deep-sea fishermen resent the thresher. It goes in after netted fish, gets tangled up and often ruins the net.

The thresher is usually black or other dark colours such as blue on top and mottled white underneath. They are the easiest species to spot because of their great long tails. Some threshers have been as long as 6 metres.

Threshers often encircle divers but they don't even attempt to turn on their tail-wagging performance.

TIGER (*Galeocerdo cuvier*)

One of the three proven consumers of human flesh, it has been blamed more than any other shark for unprovoked attacks on humans around the Australian coast. One reason is the human remains found inside a number of tiger sharks. The shark in the Coogee Aquarium that figured in the Shark Arm Murder was a tiger.

The tiger shark is often called the 'scavenger shark.' An incredible variety of objects have been found in its stomach, including a woman's handbag containing a watch which was still going. In January 1948 a native tom-tom was found in a tiger. A bottle of old and excellent Madeira wine was found in another. It prowls rivers, estuaries, harbours.

One of the most celebrated shark stories, proved in historic records, almost certainly featured a tiger shark as the hero. In 1799 a Yankee privateer was chased by a British man-o'-war in the Caribbean Sea. Finding escape impossible the skipper tossed his ship's papers overboard. The privateer was captured and escorted into Port Royal, Jamaica. The captain was put on trial for his life. As there was no documentary evidence against him he was about to be discharged, when another British vessel sailed into port. The captain of this ship reported that off the coast of Haiti a shark had been hooked. When opened up, the privateer's papers were found in the stomach. The papers were brought into court and solely on the evidence which they contained, the captain and crew of the privateer were condemned. The original papers were preserved and placed on exhibition in the Institute of Jamaica in Kingston, where the 'shark papers' as they were called, have always been of great interest.

The tiger is so named, not so much for its known ferocity but for the striped markings on its body which actually fade with age. It is generally a greyish hue, bluey-grey or brownish-grey. Its snout is short, blunt, rounded. The tiger shark is known to reach anything up to 5.5 metres in length. Its large black eye is protected by a third eyelid, or nictitating membrane, a white lid which is pulled up across the eye when it bites.

Unlike the white shark and the mako, the tiger shark likes company. It often travels with another or several other sharks. It moves sluggishly but can streak forward in an instant.

A shark's teeth is one of the best indicators of its species. In the case of the tiger shark the teeth are cockscomb-shape, with serrated edges. It often uses these on other sharks, including its own species.

The tiger shark has been responsible for a number of attacks off the Florida coast. In some instances it has charged boats. The most notorious attack by a tiger shark anywhere in the world occurred at Coolangatta, Queensland, on 23 October 1937. Two young men died. The day following the attack a 3.5 metre tiger shark was caught. Among the human remains in its stomach was the hand of one of the victims, identified by a scar. This is the only known instance in Australia where two lives were lost in one attack.

WHALER (*Carcharinus ssp.*)
BRONZE WHALER (*Carcharinus brachyurus*)
The common whaler, also known as a bull shark, along with its near relative, the bronze whaler, are responsible for the majority of attacks on humans. Like the tiger, the bulky whaler is a born scavenger. It weaves around wharves, hovers inside harbours. Whalers particularly like to bask in the warm shallows. There is an abundance of them in Australian waters. Because common whalers travel upstream into estuaries and are found in fresh water great distances from the sea, they are sometimes known as river sharks.

The whaler's body is dark grey, shading to a dull white on the belly. Teeth are serrated triangles. It has a rounded snout and small eyes. It can grow to over 5 metres in length.

A *whale shark which was washed onto a Queensland beach. They can grow to 18 metres in length.*

The bronze whaler, which grows to about 3 metres long, is found only in the clear open sea of coastal regions. It is the colour of polished bronze, fading to white below.

Unlike many other sharks the whaler tends to strike its victim only once, then break off the encounter. It may be that the unfamiliar taste of human blood makes the shark cautious. Unfortunately after that one test bite it may be too late for the victim.

In the summer of 1912 thousands of people bathed in the Narrabeen Lakes. Later they discovered a whaler, 3.6 metres, had been in there with them all along. Fortunately there were lots of fish in the lake. The whaler wasn't particularly hungry. Had there been few fish about and if a swimmer had jumped in right under his nose, it may have been a different story.

Variations in production of their young is greatest in the whaler. One caught off Sydney Heads on 22 August 1936 had 18 young, half males, half females measuring up to 710 millimetres.

The whaler got its name from the days of the whaling industry in Twofold Bay, NSW in the 1840s where they constantly attacked whale carcasses being towed into port.

WHALE SHARK (*Rhiniodon typus*)
The whale shark has nothing whatsoever to do with the whaler. Typical of the paradox of the ocean the largest shark of them all, the whale, which grows to a length of around 15 metres is the most harmless to man. A gentle giant, it will tolerate skindivers taking a ride on its back without showing either fear or irritation. This huge, 'amiable,' toothless plankton-feeder is the largest living fish. Whales run larger, but are of course, not fish but mammals.

GREAT WHITE SHARK (*Carcharodon carcharias*)
Known as the great white shark, white pointer or white death, the white shark is the classic killer, the one that all the experts agree is the most dangerous to humans. Theo Brown says it is 'the only shark that appears to be capable of a basic and limited reasoning ... it will retaliate immediately if molested or injured ... other sharks usually show a desire to escape.'

In his first encounter with this sea monster experienced underwater explorer Captain Jacques-Yves Cousteau was 'galvanized with ice-cold terror ... What we saw made us feel that naked men really do not belong under the sea.'

The subject of the length of white sharks is surrounded by

myth and controversy. Claims of 11 metre sharks have not been substantiated. In 1984 a fisheries officer lassoed a female off the coast of WA which measured 5 metres. This is one of the largest ever reliably recorded, although newspaper reports often mention white pointers 5.5 or 6 metres long.

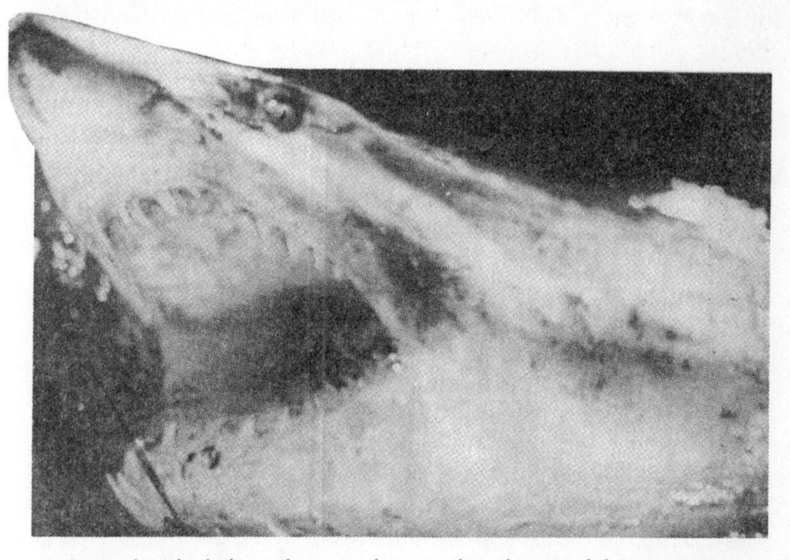

Great white sharks have a large population in the cool waters of the Tasman Sea.

Fortunately it prefers the wide ocean to the coastline and is usually found in cold water, especially where there are large populations of seals or sea-lions. However, whites have been found in tropical waters as far north as Mackay in northern Queensland.

An 18th century writer said of the great white shark, 'It is the dread of sailors in all hot climates, where it constantly attends the ship in expectation of what may drop overboard. The man that has this misfortune perishes without redemption. They have been seen to dart at him like gudgeons to a worm.'

The names given to this shark are somewhat misleading. Actually it is only whitish on its underbelly. The colour varies from a slatey grey to almost black. The dark upper and the pale lower colour is clearly defined.

The white shark has a streamlined body with a sharply-pointed head and a large crescent shaped tail of which the upper and lower lobes are roughly equal in size. The large sharp-pointed first dorsal fin is followed by a tiny second. It has a large black eye and triangular serrated teeth which are disproportionately large, sometimes measuring up to 5 centimetres.

Stories of a giant species of the great white shark which still exists today have been given some credence by scientists but none have been substantiated. The following story is cited in David Stead's *Sharks and Rays of Australian Seas* published in 1963.

In 1918 an incident so affected the seasoned cray fishermen of Port Stephens, north of Newcastle, they refused for several days to go out to their customary deep water grounds off Broughton Island. It seems that one afternoon the living daylights had been scared out of them, when a shark of the most awesome proportions joined the party. This startling monster, its terrible head at least as large 'as the roof of the wharf shed at Nelsons Bay,' began swallowing their crayfish pots; some as much as a metre in diameter containing as many as three dozen fair-sized crays. The creature even took the mooring lines. In their time these men had seen huge sharks and even bigger whales, but never anything like this. They estimated the shark to be 35 metres long. The local fisheries inspector accepted and recorded their estimation.

Even this mysterious giant would be small compared to the giant fossil shark whose teeth up to ten times larger have been found in inland places some kilometres from the ocean.

A reconstruction of jaws of a giant shark that lived on this planet in the dawn of time has been reconstructed at the American Museum of Natural History in New York. Six men can stand in its yawning mouth. Small wonder legend claims the great white shark, not the whale, was the ocean colossus that swallowed Jonah.

Another rare piece of evidence of the existence of a giant species of the white shark comes from another unsubstantiated story told to diver and shark repellent researcher Theo Brown by the skipper of a charter fishing boat off the Queensland coast in 1960.

The skipper said he was on the bridge making a sweep with his binoculars when suddenly he focused on a massive, grey bulk moving slowly up from the depths. He gasped with amazement. It was an enormous fish with huge fins and a tail bigger than the tail of any whale he had ever seen.

He yelled to the crew that there was a whale off the starboard quarter. He raised his binoculars again and shouted, 'My God. It's not a whale its a bloody shark, a giant shark!'

The wide-eyed crew saw the shark move in under their boat where it hovered in its shadow a full 15 minutes. They agreed that, from stem to stern, the monster shark was the length of their boat, 26 metres. The whole incident was quite preposterous. Because of this and for fear of being laughed at, they made an agreement not to make their story public.

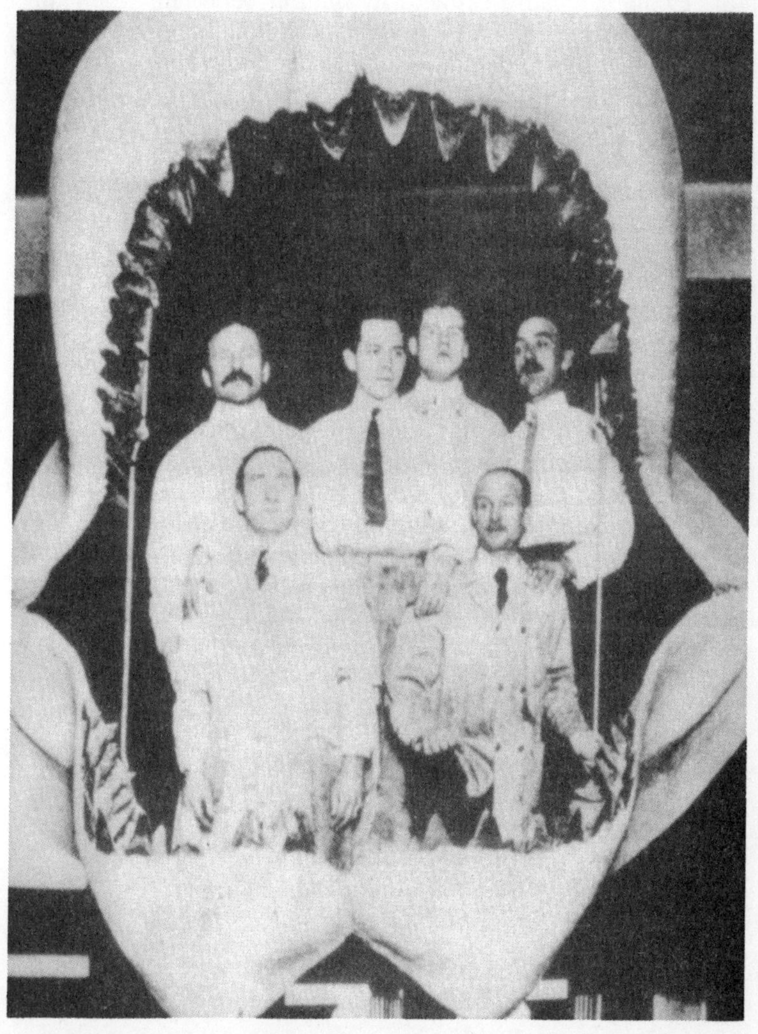

The jaws of a prehistoric shark reconstructed from fossils at New York's American Museum *of* Natural History.

WOBBEGONG (*Orectolobus ssp.*)
The wobbegong is sometimes called a carpet shark because of its pattern of blotched and marbled colours. There are a number of species found in Australian waters.

Unlike the grey nurse, the 'wobby' shows little fear of humans, so it is just as well it doesn't eat them. Nevertheless the wobbegong can cause some nasty injuries especially when cornered. One species, which ranges in New Guinea waters, has caused a number of deaths among native swimmers, when bite wounds have turned poisonous.

Wobbegongs are usually found lying on the sea floor in schools. They like the shallows. There are numbers of them in Sydney Harbour. They are also found on reefs where their multi-patterned bodies blend nicely with the environment. Heaven help a skindiver who accidentally treads on one!

The wobbegong is a curious looking creature. It has a flat, wide body and a weird mouth surrounded by a fringe of skinny flaps. Its pointed teeth are long and sharp. It is sandy in colour with darker-brown mottled blotches. It grows to lengths between one and 3.2 metres. One species grows only to 50 centimetres. However, giant wobbegongs, 4.8 metres have been known.

It was one of these giants, attracted by the convulsions of a speared groper, that flashed out of an underwater cave and made straight for spearfisherman, Rex Gallagher of Shellharbour, NSW early in 1953. It charged his face mask, seemingly attracted by the bright metal band. The mask was ripped aside; the snorkel snapped. Gallagher was slashed about the face, but he managed to get away.

CHAPTER FIFTEEN

Repellents

It is a good-to-be-alive morning. The sun beckons from a metallic blue sky. The air shimmers with bright promise. Along the chain of biscuit sands that fringe that eastern extremity of sprawling suburbia, the multicoloured umbrellas pop up like mushrooms.

In 10,000 homes sandwiches are sliced. Chilled bottles drop into ice containers. Towels and surf boards are flung into car boots. Children call, impatient to be off. It is a summer Sunday morning in Sydney, Australia. The exodus to the ocean has begun.

Seventy percent of our population live on or near the coast and the story is being repeated in towns and cities around the continent. Somewhere deep in the subconscious is a lurking fear. But memories are short and the water is far too inviting.

Along the main beaches where sharks are prevalent, the defence system is in operation. The alarm bells and shark sirens are silent and waiting. Lookouts and spotters scan the waves, the shark plane buzzes overhead. Out there beyond the furthest line of breakers is the best defence method to date: the meshing nets.

The only 100% birth control method is total abstinence. The only 100% protection from shark attack is to remain on dry land. For hundreds of thousands of Australians this is like asking to give up ... well, their car.

The repellent known as 'Shark Chaser' became as important a part of a survival kit as the life jacket. It was never proved to be shark-proof but the name helped — psychologically at least. The experiments with 'Shark Chaser' began when it was found that while sharks were not averse to eating other sharks, when they were alive or only recently dead, they had a positive distaste for shark flesh that had decomposed. The closest scientists came to rotting shark was a copper acetate. This was mixed with certain poisons and standardised into packets which were distributed to the forces in the Pacific war theatre. It was claimed, by those who created it, to be 95.2% effective.

Pearl divers had long had an inkling of the shark's aversion for rotting shark and in some areas the pearling season opened with a shark hunt after which the dead sharks were left to decompose at the bottom of the sea.

The real effectiveness of 'Shark Chaser' was never proved. In fact more than one survivor reported the shark appeared to swallow the stuff without a murmur. Another wartime repellent, devised long before underarm deodorants, were the chlorinated Halazone tablets to camouflage body odour. Experiments with captive sharks showed they did not like the 'swimming pool' taste of the chlorine. These capsules also became part of the survival kit.

Imagination has no bounds when Australians begin making suggestions about ways to ward off shark attacks. These have ranged from underwater hooters to tingling bells on swimmers' wrists. The whole paraphernalia of modern warfare has even been proposed and tried: explosives, rifle fire, aerial machine-gunning.

In 1958 South African destroyers dropped depth charges following two fatal attacks within two days of each other. But such all-out methods are part of the hysteria that sometimes follows multiple attacks.

One method proposed in a letter to a Sydney newspaper was taken up and improvised upon overseas. The idea was for a pipeline to be attached to a running motor car exhaust which pumped a 'wall' of carbon dioxide bubbles from the floor of the sea. In New Jersey a perforated hose was laid beneath the sea between two jetties. At first the sharks backed off before the curtain of bubbles. But they got used to them and were soon passing back and forth freely. Electrical barriers have also been tried. They proved costly and inefficient although the sharks did find the volts rather revolting.

The first positive shark repellents came into being in the early war years. Scientists were hard pressed to come up with a suitable repellent. In 1956 a US oceanographer research party fastened a packet repellent, similar to 'Shark Chaser', to a dead fish and dangled it over the stern of a launch. A shark darted through the cloud repellent and swallowed the fish. It circled, then came in again, this time swallowing the repellent, but promptly spat it out. Captain Cousteau once dissolved five copper-acetate tablets in the face of a circling white shark off the Cape Verde Islands. It responded by charging him.

In his book Sharks: A Search for a Repellent (1973), Theo Brown, who tried to save the life of Ken Murray in Middle Harbour in 1960, tells how he tried out the US Navy's 'Shark Chaser' for himself. The 'new improved' 'Chaser' contained one part copper acetate to four parts of negrosine-type dye, the dye's purpose being the same as that exuded by the octopus for defence. The location for the experiment was 112 kilometres north-east of Townsville. Several 44-gallon drums of blood and offal from the Townsville meatworks were poured into the sea. Larger baits were strung to the boat. In minutes, dorsal fins were cutting the water. Larger sharks joined the circling promenade around the boat, the fins coming in closer. As they did so the movement of the fins became erratic, the sharks more agitated. Suddenly the boat rocked as a large shark thumped against it in a grab for the hunk of beef bait. A bag of the 'Shark Chaser' was split and emptied over the shark. The black repellent spread and the shark churned the water violently, the beef in its jaws. Already there were scores of sharks surrounding the boat and suddenly the sea exploded as the frenzied sharks fought over the bait. The crew tipped every ounce of the repellent into the sea and the boat rocked as the sharks smashed against the hull. Now the sharks were slashing and biting at each other in the boiling black and red cauldron of the sea, while those on the boat were awe-struck by this almost supernatural display.

Captain Cousteau had these comments about advice from 'experts' on how to scare off sharks:
Lifesaver: 'Gesticulate wildly.'
Cousteau 'We flailed our arms. The grey shark did not falter.'
Helmet diver: 'Give 'em a flood of bubbles.'
Cousteau: 'He released a heavy exhalation. The shark did not react.'
Hans Hass: 'Shout as loud as you can.'
Cousteau: 'We hooted until our voices cracked. The shark appeared deaf.'
Air Force Officer: 'Copper acetate tables fastened to leg and belt will keep sharks away if you go into the drink.'
Cousteau: 'Our friend swam through copper-stained water

without a wink. His cold tranquil eye appraised us. He seemed to know what he wanted, and he was in no hurry.'

In the end Cousteau and his team used billy sticks. 'After seeing sharks unshaken with harpoon through the head, deep spear gashes on the body and even after sharp explosions near their brains, we place no reliance on knives as defensive arms. We believe better protection is our "shark billy," a stout wooden staff, 4 feet long, studded with nail tips at the business end. The nails keep the billy from sliding off the slippery leather, but do not penetrate far enough to irritate (and thus arouse) the animal.'

Sound may yet prove to be the ultimate shark deterrent. These are the lines along which experts like Theo Brown have worked. Already sounds have been developed which repel certain species of shark. 'Now we have to refine the signal pattern so that the finished sound will effectively repel maneaters world-wide.'

An incident in World War II is an example where sound may have saved the life of a crash survivor. A tiny raft with a pilot inside was being used like a volley ball between several sharks. The helpless and terrified airman tried running his fingers along the side of the raft. Miraculously, the high, squeaky note seemed to put the marine players off their game. They suddenly disappeared, although they may have done so for other reasons.

Theo Brown believes that one day underwater sonic waves may clear places like Port Jackson (Sydney Harbour) of all sharks. More immediately, it has been suggested that sonics built into the fuselage of planes and activated by water pressure, may protect crash survivors.

Pearl divers use the 'underwater scream' as a defence against an unwelcome shark. Sri Lankans cry: 'Mora! Mora!' Dr Hans Hass, author of Diving to Adventure, was a foremost exponent of the scream theory. It had succeeded with sharks in the Mediterranean. Sharks in Australian waters seemed to be more tone deaf. When he visited Australia Dr Hass noted that the sharks he encountered here were less hesitant and wary of humans than those of the Caribbean or Red Sea, sometimes coming so close they had to be held off with spears. Actually it is the vibrations caused by the scream that are more likely to irritate the shark, just like a loud yell close up to a human ear. Others say it is the bubbles caused by the scream. Sharks don't like bubbles.

It took 15 years, from the time surfing caught on in Australia after World War I until 1934, a grim year for sharks, for the NSW government to act against shark attack. The Shark Menace Advisory Committee was appointed that year, under the chairmanship of Adrian Curlewis, to find ways of combating the shark. The shark alarms and the look-out towers were already instituted. Various clubs had their surfboats equipped with harpoons. An aerial shark patrol was already operational. A Dragon DH4 winged over the metropolitan beaches bellowing 'All clear' through loudspeakers. None of these precautions stopped the fatalities. In the early months the committee received more than 100 proposals from the public. Eventually it was agreed that meshing seemed to be the most effective and the most logical answer.

Of course there were the cynics. 'A futile waste of public monies,' they called it. They caused some delays. In the period February 1935 to February 1937 there were five more deaths from shark attack in the Sydney and Newcastle areas. Earlier meshing might have saved them. Meshing proved a smash success, even though it was never claimed to be 100% shark-proof.

Heavy-gauge nets, approximately 25 by 25 centimetres, are submerged at intervals from buoys to the seaward side of the breaking wave. They are suspended between a float line on the surface and a lead line on the bottom and anchored at each end. Although sharks can swim over or around the mesh they are usually unaware of its presence and tend to swim into it, becoming entangled. The mesh interferes with their gill function and as they struggle and become more entangled they normally die from suffocation.

After about 24 hours the mesh is retrieved and those sharks still alive are killed. Thus the local population of sharks decreases, the remaining sharks have access to a greater share of the available food supply and fewer hungry sharks remain to attack humans.

From October 1937, when the first nets were laid, the attacks at Sydney's surfing beaches ceased, apart from two attacks in quick succession early in 1940. North Brighton Beach in Botany Bay was not meshed.

It will be seen that shallow water is no safeguard. One bizarre incident early in World War II emphasised this. An Indian Army ambulance driver was washing his vehicle in a shallow river in Iran. The water was less than knee deep. A shark knocked him off his balance and chewed his leg so badly he bled to death.

There have been incidents reported in this book in which the shark practically came on to dry land. In 1963 Bill Harvey and aptly-named fishing mate, Bert Fisher, were digging for worms on a mudflat in the Georges River. There was a loud splashing behind them. They turned and almost jumped out of their skin. A 1.5 metre shark had beached on the bank several feet behind them. It must have been scooting after a school of mullet. With a few frantic twists of its tail and body, it jerked itself around and slithered back into the water. The lads took off from the sandbank in a great hurry.

CHAPTER SIXTEEN

▲▲▲▲▲▲▲▲▲▲▲▲▲▲▲▲▲

Moment of truth

If the worst should happen, if you are confronted by a menacing shark, remain cool, if that is possible. In his handbook, Sharks (1975), Peter Goadby suggests keeping your eye fixed on the shark. It's your display of 'confidence.' It shows you have control. Remember, if you have seen the shark first, at least half the danger has been eliminated.

A swimmer without underwater weapons has only his arms and legs for protection. There have been times when a good hard blow, landed in the right spot, has actually stopped the shark in its tracks.

The old journalistic saw 'Man Bites Dog' has been used on the shark. One story out of the Solomon Islands, tells of a fisherman, Elison Sevo, who bit a shark on the nose after it snapped at his leg. The shark took off smartly. A hard bite on the snout, however, did not persuade the shark to release Ray Short's leg at Coledale, NSW in 1966.

There is one other weapon a person can use. Although the shark may be infinitely superior in its own environment, the human brain is four times larger. There is always a possibility the shark can be outwitted long enough to buy time until help comes.

One school of thought believes a sudden noise or movement will scare the shark away. Another group of experts believe abrupt movements could be dangerous. Certainly, it is best to move a little so that the shark gets the message you are alive and well and not merely a corpse for easy picking. If you are cut, stop the bleeding in any way — quickly. If there is more than one of you in the sea and the shark threatens, gather close together for all-round observation.

Landing in the water and getting out are the moments of maximum danger from lurking sharks in rivers and harbours. One should be watchful although even the keenest observer could be fooled. Underwater, one always keeps a sharp lookout. Stay alert, but not jittery. At best these suggestions will reduce the danger of attack. There is a no 100% safeguard.

Lying prone is strongly recommended by those who should know. It usually brings the shark to the surface where you can see it. It is less tempting than a pair of legs dangling, like an inviting bunch of bananas below the surface.

A relaxed breast stroke is the safest to use when there's a shark in the neighbourhood. Free style, backstroke, butterfly or dog paddle could excite enough curiosity to bring the shark in to investigate.

People on rafts or boats should dispose of fish, food and body waste after checking there is no shark hovering close by. No sense in asking for trouble. If there is a shark in the vicinity, and someone is seasick, it is wiser to vomit inside the boat.

Experienced deep sea fishermen know that when a line with a hooked fish on the end goes dead it is best to leave it for a few minutes. A shark may have taken a piece out of the catch and if it is jerked away might well come right up practically into the boat after it!

TREATMENT

The flesh is weak and physically humans are entirely exposed to injury. A person with superb physique, as many divers and surfers are, is as vulnerable as an infant child in the voracious jaws of a killer. Shock is a contributing factor to death, second only to loss of blood.

David Webster in his Myth and Maneater (1967) points out that native divers are shark-orientated. Pearl divers are psychologically attuned to the alien underwater world where they spend half their lives. Metropolitan people are at one moment alive in a sparkling world of surf and sun, in the next shattered by the onrush of giant fish whose rending teeth produce paroxysms of pain and exploding blood vessels. The shock alone can kill them and it often does.

Shark victims must be treated as soon as they are brought ashore. The first call for medical assistance should be to the Red Cross Blood Transfusion Service; the next to the ambulance. The victim should be laid head down to relieve bleeding. A local doctor should be summoned to give morphine injections to deaden the pain. On the way to the scene the blood bank vehicle could pick up a doctor and give the blood transfusion on the spot. The doctor will also give antibiotics to reduce the very real danger of infection, the reason why many shark attack patients have died.

Beach bags containing blood serum and morphine have long been standard in the first aid room of South African beach clubhouses. Durban general practitioner, Dr Ian Dalziel, who had treated many shark victims, had some definite views when he arrived in Australia in 1967 with the South African surf team. He said victims stood a better chance of recovery if they were not covered by a blanket. They should be left still on the beach at least half an hour. He said the ambulance should be driven slowly; no more than 30 kilometres per hour. It was important the patient be treated with the head down and morphine and fluids be given intravenously. 'All of the shark victims treated in this manner have lived while those who have not received this kind of treatment have died.'

This brought quick response from Dr Gordon Archer of the NSW Red Cross Blood Transfusion Service. He said this method had been in operation in Australia for two years. He was aware that a blanket would make the victim sweat and the blood would go to the skin. Less blood would reach the brain and the victim would suffer from even greater shock. 'The practice in the past was to rush victims to hospital, at high speeds,' said Dr Archer. 'Ambulances now carry blood plasma and blood transfusion equipment.'

When a fatal attack makes the headlines, especially if it occurred 'close to home,' we start taking precautions again. But as time passes we forget. We swim outside the shark enclosure. We stray outside the flags on the surfing beaches. We swim, a little further out, each time. And then it happens.

'After a time, no doubt young men will again take risks which young men always run,' said an editorial in the Sydney Morning Herald following a shark tragedy in Middle Harbour. 'But parents have a responsibility for young children, and it should not require a bloody sacrifice in the summer sea to remind them of it.'

We, who prey without remorse on all forms of life, regard as unnatural and abhorrent those few creatures that prey on us. But the creature we probably fear most, more than the pouncing leopard, the stalking tiger or the poisonous snake, is the shark. We meet the shark in its own domain and, if we are weaponless, under its own terms.

A letter which appeared in the Sydney Daily Mirror, 27

November 1967 written by R. D. Finlayson of Mosman, NSW sums it up this way:

'The motor car is a far more frequent and dangerous killer and can cause painful and horrible deaths. But are we terrified out of our wits at the sight of the motor car? Of course not. We exercise caution and care to lessen the risks of accidents. Why not do the same with sharks? I hate to see them slaughtered and hung upon a stick to be looked at with fear and hatred. Leave these graceful, beautiful creatures alone in the sea where they belong, and keep out of their way.'

This view was echoed in June 1993 by Dr John Stevens, a CSIRO shark biologist in Hobart, following the hunt for the white pointer which killed Therese Cartwright.

'I don't understand this paranoia about shark attacks ... people don't go around blowing up vehicles if there is a car accident.'

A question arose in the NSW Legislative Assembly on 24 November 1967 regarding a headline spread across the front page of the Sydney Daily Mirror the previous day. The headline read: MANEATER: 37 DEAD AND MAIMED. Inside the paper the figure referred to the number of crippling shark attacks in Australia that took place in the period 1919 to 1963. The Leader of the House agreed that 'the headline is likely to give overseas visitors to our beautiful beaches a totally wrong impression.'

The report of attacks in Australian waters, concentrated into a single volume such as this, is likely to give an equally false impression. Compared with the number of people and the number of hours spent in the water the record of shark attack on humans is infinitely small. As long as there are sharks in the sea there will always be the remotest possibility.

The lesson to be learned is to try and improve on that million to one chance by exercising caution, by resisting the temptation to swim in unprotected waters, however inviting they may be on a hot day, and by obeying the signs and the rules.

We might also seek to be aware of our own influence on the delicate ecology of the marine environment, of the fact that the dangers of shark attack in any area may be intensified when their food supply is reduced by excessive fishing or other factors.

The rest is left to fate.

SELECTED BIBLIOGRAPHY

Budker, Paul. *The Life of Sharks*, Columbia University Press, USA (1971)

Brown, Theo. *Sharks — The Search for a Repellent*, Angus and Robertson, Australia (1973)

Butler, Jean. *Danger Shark!* Robert Hale, UK (1965)

Claiborne, Robert. *On Every Side the Sea*, American Heritage Press (1971)

Cook, J.J. *Nightmare World of Sharks*, Dodd, Mead, USA (1968)

Coppleson, Victor. *Shark Attack; How, Why, When and Where Sharks Attack Humans*, Angus and Robertson, Australia (1988)

Cousteau, Jacques-Yves. *Shark: Splendid Savage of the Sea*, Cassell, UK (1970)

Cropp, Ben. *Shark Hunters*, Rigby, Australia (1964)

Davies, David. *Shark*, Routledge & Kegan Paul, UK (1964)

Edwards, Hugh. *Sharks and Shipwrecks*, Lansdowne, Australia (1975)

Ellis, Richard & McCosker, John E. *Great White Shark*, Harper Collins, USA (1991)

Gilbert, Perry. *Sharks and Survival*, D. Heath, USA (1963)

Goadby, Peter. *Sharks*, Ure Smith, Australia (1975)

Grant, Ern. *Fishes of Australia*, E.M. Grant P/L, Brisbane (1987)

Helm, Thomas. *Shark!* Robert Hales, UK (1962)

Lineaweaver, Thomas. *Natural History of Sharks*, Andre Deutsch, UK (1970)

McCormick, Harold. *Shadows in the Sea*, Chilton Books, USA (1963)

Prosperi, Francis. *Lord of the Sharks*, Hutchinson, UK (1955)

Riedman, Sarah. *Focus on Sharks*, Abelard Schuman, USA (1969)

Stead, David G. *Sharks and Rays of Australian Seas*, Angus and Robertson (1963)

Webster, David. *Myth and Man Eater*, Peter Davies, UK (1967)

Whitley, Gilbert P. *Fishes of Australia*, Aust. & NZ Royal Zoological Soc. (1940)

PERIODICALS

The Victorian Naturalist, April 1889

The Medical Journal of Australia, 31 July 1920, 15 April 1933

National Geographic, February 1968

Scientific American, July 1962

NEWSPAPERS

Sydney Morning Herald
Melbourne Age
Brisbane Courier Mail
Adelaide Advertiser
West Australian
Telegraph Mirror
Illawarra Mercury
Hobart Mercury

INDEX